LAST PLACES

BOOKS BY LAWRENCE MILLMAN

Our Like Will Not Be There Again

Hero Jesse

Smell of Earth and Clay

A Kayak Full of Ghosts

The Wrong Handed Man

Last Places

An Evening Among Headhunters: And Other Reports from Roads Less Traveled

Northern Latitudes

LAST PLACES

A Journey in the North

Lawrence Millman

INTRODUCTION BY

Paul Theroux

A Mariner Book
HOUGHTON MIFFLIN COMPANY
BOSTON • NEW YORK

First Mariner Books edition 2000

Visit our Web site: www.houghtonmifflinbooks.com.

Library of Congress Cataloging-in-Publication Data
Millman, Lawrence.
Last places: a journey in the north / Lawrence Millman.
p. cm.
Originally published: 1990.
"A mariner book."
ISBN 0-618-08248-4
1. Millman, Lawrence—Journeys. 2. Voyages and travels. 3. North Atlantic Ocean. I. Title.
G470 .M52 2000
910'.91631—dc21 00-046548

Printed in the United States of America

Book design by Robert Overholtzer

QUM 10 9 8 7 6 5 4 3 2

CONTENTS

Improvement makes straight roads; but the crooked roads without improvement are the roads of genius.

—William Blake

Ah! These commercial interests—spoiling the finest life under the sun. Why must the sea be used for trade—and for war as well? . . . It would have been so much nicer just to sail about, with here and there a port and a bit of land to stretch one's legs on, buy a few books and get a change of cooking for a while.

—Joseph Conrad,
"An Outpost of Fortune"

They went ashore and looked about them. The weather was fine. There was dew on the grass, and the first thing they did was to get some of it on their hands, and put it to their lips, and to them it seemed the sweetest thing they had ever tasted.

—*Vinland Sagas*

INTRODUCTION

THIS IS A BOOK by a traveler—part ethnographer, part vagabond—who is both an inspired talker and a great listener. He is also the most independent of souls. *Last Places* is full of odd stories and keen observations. In Lawrence Millman's telling, the gruesome uniqueness of a ritual whale slaughter, the Faeroese *grindadrap* or whale hunt, is shocking and insightful. As for the author's independence, two hundred pages into the book and many miles into his travels (he is in northern Newfoundland), Millman is flummoxed and rents a car. He apologizes to the reader for doing so this single instance. Most of the time he is hiking or onboard ship in places where few travelers ever go and even fewer have written about. Ten years after the book's publication, these destinations still count as last places.

Millman's stated intention, a characteristic traveler's conceit, is his following the Viking route from Norway to the New World, Bergen to Newfoundland. He comes to rest in L'Anse aux Meadows, which he reminds us is a corruption of L'Anse aux Meduse—Jellyfish Creek—perhaps the site of the Vinland of the Sagas. In another place he writes more candidly and with his usual gusto of his travel on this long, mostly oceanic journey being "a celebration of the North Atlantic wilderness."

He sets out his notion of travel in an early paragraph in the book:

> Travel is the realm of the improbable adventure, the quick fix, the ship passing in the night. It entitles you to meet interesting people whom you otherwise would never meet even if you laid traps or advertised for them. Not only do you meet them, but you also *unmeet* them, all in the space of, it often seems, a mere compacted evening. As there is so little time, bodies in motion tend to drop their guard and immediately get on with their stories. Then the proverbial ships part, each to its own destination, never again to brush each other's wake.

This experience is familiar to me. It has been my lot to have spent a large part of my own traveling life pursuing chance encounters—one night stands, in the chastest sense of the expression. I can also identify with Millman's love of being off the map and his liking for unusual facts, even ones not related to his present journey. In *Last Places* we are treated to Tibetan and Arab proverbs, the notion of "muckle fever," Custer's last words, and the scientist J.B.S. Haldane's comment on the mentality of the Creator: "He has an inordinate fondness for beetles." While not writing with particular intimacy about himself, except where it bears upon his journey, Millman grimly rejoices in describing his bout of seasickness and an episode of his becoming lost. But usually he has a stomach of iron and a wonderful sense of direction.

As well as willingness, humility, and optimism, a traveler needs the patience of a saint. Millman is game—quietly anatomizing in the gory confusion of the whale slaughter, receptive to the grimmest hospitality, resourceful in dealing with drunks (the northern latitudes are absolutely heaving with boozers). The omnivorous natives eagerly share their entrees with Millman. He eats mataq—"the black and white mottled skin and first layer of blubber of the narwhal"—raw, from a fresh kill. He fends off the Greenlandic wife swappers—most days in

Greenland being Sadie Hawkins Day; and he finds something to like, something of value, just about everywhere he goes in this unfrequented part of the world. Perhaps this is not surprising, since so many of the people he meets are local versions of himself, self-sufficient loners who love awful weather and disgusting meals.

Because Millman is an appreciative listener, and a note taker, interested in local myths and etymologies, this is, among other things, a collection of local stories. Stories in Foula and the Faeroes, stories in Greenland and Labrador and Iceland; stories in Hornbjarg, where the lighthouse keeper at "the lighthouse at the end of the world" has a library of sixteen thousand books; stories about animals and fish and demigods; a story about Eric the Red's father, a murderer who was rusticated to Hornstrandir. Some of the stories are long, some very short.

In the small Greenland settlement of Qaqortoq, an Inuit tells him a story about a raven and a seagull, and because this is *Last Places,* Millman has met the man purely by chance on a fish dock, and the man is a seal hunter who, as he relates the story, also happens to be "flensing a ringed seal." The story goes:

> Once a raven and a seagull got into a fight over a piece of meat. The raven was on the Inuit's side, the seagull on the side of White Man. They fought for days, for weeks, even for months. Whoever won, his side would be the stronger. They tore and bit furiously at each other. At last the seagull won: White Man would be stronger and more plentiful than the Inuit. But by the time he flew away with his piece of meat, it had become quite rancid.

Like this story, *Last Places* is also wise, humane, funny, and, in all senses, peculiar.

PAUL THEROUX
July 2000

Chapter 1

EMBARKATION

ONCE UPON A MUCH EARLIER time I was traveling on a Turkish Maritime Lines ferry from Istanbul to Trabzon, city of nightingales, golden scimitars, and mysterious jails capable of swallowing a man forever. Shortly after Samsun, a northeasterly storm hit and turned the Black Sea into a boiling mountain range of waves with dispositions scarcely less violent than genuine ocean waves. Our boat bucked and rolled, pitched and corkscrewed with ever-growing determination, as if preparing itself for a capsize. Down below I bucked and rolled in an E-class bunk. Defrocked civil servants and indigent goatherds travel E class — high tiers of bunks in the bowels of a starboard that takes every knock from the sea like a hug from a long-lost brother. I lay at the bottom of one of these tiers, a position that began to disturb me when a plague of seasickness deprived first one passenger, then another, of his supper. Nobody bothered to get up; they simply leaned over the side and heaved. I backed up against the wall and tried hard to think of certain down-home Yankee mundanities, like Boston ward heeling or the CIA's attempts to dissolve Castro's beard.

And yet a quite pleasant mood prevailed among my fifty or so bunk mates. They'd all seen much worse than a little half-digested sheep's brain salad on the floor. For the E-class Turk, life is mean, oppressed, and mercifully short. Seasickness, by contrast, is festive. Someone was drunk and singing about (I was told) a famous Anatolian *femme fatale*. An old man in a fez and blue-striped pajamas reached down and offered me the stem of his *tchibouque;* I took in a bitter flavor not unlike qat leaf cut with overworked gastric juices. The old man grinned. It wasn't his grin, though. Leprosy had drawn back his upper lip and forced his mouth into a permanent facsimile of a grin.

Suddenly a peddler came down the aisle hawking a tray of sizzling corn on the cob. Now everything, from the flavor of the *tchibouque* to the bittersweet aroma of underarm and foot to the rising tide on the floor, began to take on a quality of horror. I hadn't been seasick yet, but I knew that if I stayed put, the corn would ignite the fuse to my stomach in seconds. I grabbed my pack and dashed down the corridor to the ship's bar. There drunk Turks might be sticking knives into each other, but at least there wouldn't be any corn.

"Pardon me," I said to the bartender, who was drying a row of *raki* glasses, "but I wonder if I could have a shakedown here for the night?" "No Inglis screeched," he said, but gestured me toward a bench where three of my fellow passengers were already seated.

On this bench was a man whom I took to be an American, puking noisily into a plastic bag. His face suggested Lazarus a few hours before the legendary meeting with Christ. Beside him was a girl whose head was buried in her hands and who kept opening and closing her legs quickly, as if beaming messages at sea. The third person was a raw-boned man of around thirty-five, with hair the color of wood shavings. As he appeared to be the most composed of the three, I sat down next to him. Indeed, he was so composed that when a particularly violent lurch sent two dozen *raki* glasses crashing to the floor, he didn't bother to look up from the notebook in which he was writing.

A short while later I found out that the man's name was

Gudmundur. He was an Icelandic trawler captain, the son of another Icelandic trawler captain and grandson of a third, all from the town of Akureyri. The U.N. Food and Agriculture Organization was sending him to Calcutta to instruct the Bengalis in the finer points of fish management; he was traveling overland — over the water, too — because airplanes made him nauseous.

Are you organizing your notes? I asked him. No, he said, I am writing a rhyme about a girl I met in a bistro in Istanbul. She had hair like a raven, which is not so common in Iceland. I took her to my hotel room and we made screw. Then her boyfriend, he came to the hotel room, too. He was very unhappy. He wanted to kill us with his knife, so I had to take his knife from him. Then he wanted to make screw, too. I told him I am not a feminist. (You mean homosexual, I interjected. Yes, homosexual, he said. My English it is not so remarkable.) Now the girl was unhappy. She says why do I not like her boyfriend. I tell her I like her boyfriend very much, but I do not want to make screw with him. The boyfriend he has gotten his knife again, so I must take it away again. I have always wanted to make screw with a German, he says. I am not a German, I tell him, but an Icelander. Same thing, he says. We are going to be married, the girl says. Next month, the boyfriend says. And then the three of us sat down and drank a bottle of *raki.* I am calling the rhyme "Girl with the Raven Hair."

Now I saw he was carrying a curiously embroidered little pillbox. As if he had divined my thoughts, he slid back the lid to reveal a small sliver of charred wood. "You can imagine a piece of Huckleberry Finn's raft?" he said.

"*That's* a piece of Huckleberry Finn's raft?"

He shook his head. Not that he wouldn't be proud to own a piece of Huckleberry Finn's raft, but this was actually a piece of its Icelandic counterpart. Certainly I'd read *Njal's Saga?* This splinter came from Njal's own cowshed, which had been burned to the ground — along with Njal himself — by Flosi and his men in the year 1011. It was possible that their feud was still going on in remote parts of the country even today. He had heard of one man from Melrakkaslétta who claimed descent from Njal and who tried to push a neighbor's new Moskvitch off a cliff because the neighbor claimed descent from Flosi.

I asked Gudmundur how he had come to acquire a splinter from such a legendary cowshed. He said he'd gotten it from his uncle, who used to have a farm at Bergthorshvoll, where Njal had had *his* farm. One day the uncle was laying in some new manure and he'd found all this old char in the ground . . .

Now Gudmundur gave me this shard of ancient Icelandic life to examine. It reminded me of the fingernail of a saint I'd once seen resting in an aspirin bottle in Ravenna, Italy.

"I keep it with me for good luck," Gudmundur said. "Just like your rabbit's fits . . ."

"You mean rabbit's *feet*," I corrected him.

"Yes, that's what I mean," he said. "My English is not so remarkable." And then with his pocketknife he cut off a small sliver from this already small sliver and presented it to me as my personal, thousand-year-old good luck charm.

I was quite touched. Perfect strangers don't usually give you pieces of their national heritage. I wrapped the splinter securely in a wad of tinfoil and put it in my rucksack.

The rest of the journey, though still a bit tempestuous, passed without incident. Yet when the ferry reached Trabzon, the local police went through my effects like dogs on a gut wagon, and later I couldn't find the little splinter from Njal's cowshed. It must have been what the police were looking for. Trabzon I remember for only one other thing: a vulture, somewhat disoriented, flew through the window of my hotel room and landed on the sink with a loud flutter.

* * *

Travel is the realm of the improbable adventure, the quick fix, the ship passing in the night. It entitles you to meet interesting people whom you otherwise would never meet even if you laid traps or advertised for them. Not only do you meet them, but you also *unmeet* them, all in the space of, it often seems, a mere compacted evening. As there is so little time, bodies in motion tend to drop their guard and immediately get on with their stories. Then the proverbial ships part, each to its own destination, never again to brush each other's wake.

I never saw Gudmundur again.

Throughout the 1970s, however, I kept Iceland in mind. I cherished the image of a solitary butt of lava where everybody was happily murdering everybody else owing to the events of a millennium ago; between murders, they wrote poetry. I read the Sagas; I read W. H. Auden's *Letters from Iceland;* I even read Ida Pfeiffer's *A Journey to Iceland,* a bilious nineteenth-century travel account that reveals Iceland's huge, disabling flaw — lack of butterflies (the author was a lepidopterist). I studied maps of the island. That long-lost little splinter, I finally realized, hadn't been lost at all; the Turkish police hadn't thought it an illegal substance and hadn't confiscated it. Not at all. It was still in my possession, sticking thornlike in my side one moment, banefully in my foot the next. Then one day I knew there was only one way to get rid of it, and I hopped a flight to Iceland.

Right away the island strengthened my belief in the old idea of topographical resonance: You are what you inhabit. The Swiss are overly cautious and act as if one false step will dislodge their hard-iron pitons and send them plunging to their deaths. Australians are parched, weatherbeaten, and a little laconic, like their own outback. As for Icelanders, their behavior seemed to resemble their eccentric butt of lava. Nearly everyone I met, as befits an island that is alternately fire and ice, seemed to be marching to the beat of a distinctly different drummer. They included: fishermen who tasted the sea to determine whether it had any fish; a farmer named Magnús who knew each of his seven hundred sheep by name; a country priest who pointedly used the Icelandic patronymic, Jésus Jósephsson, for his Savior's name; the bard Sveinbjörn Beinteinsson, who incanted ancient *rimur* to rock band accompaniment; and the spirit of Mao Tse-tung, called up at a Kópavogur séance, who (which?) spoke perfect Icelandic and defended this by saying that Icelandic was the language of the afterlife. A strange admission for the chief architect of the Chinese Revolution, but nobody at that séance quarreled with it.

One of the people I met was a bush pilot named Thorsteinn who regularly flew his turboprop five hundred miles from Reykjavik to Kulusuk, East Greenland, to visit his Inuit girlfriend. The first time I accompanied Thorsteinn on one of his missions, I distinguished myself by performing an icy pirouette in cold November seas; the second time, I became violently ill from eating tainted seal meat, the same tainted cuisine, so a local hunter smilingly informed me, that had done in the Danish explorer Knud Rasmussen. Misfortune always seemed to beckon when I traveled to Greenland, yet I felt such an attraction for that big empty icebox of an island that I went to live there for six months after my finances could no longer cope with Iceland's monumental inflation.

I found myself increasingly smitten by the North. One year I was camping in the Barren Grounds of Labrador and watching three wolves, seasoned professionals, cut caribou calves from a herd. Another year I visited the Faeroe Islands and ate chunks of whale blubber that packed enough grease to lube an eighteen-wheeler. Yet another year I was back in Greenland, studiously suffering frostbitten toes.

In travel, as in food, one man's caviar is another man's soggy dumpling. The person who finds the Taj Mahal boring might find the Corn Palace in Mitchell, South Dakota, a joy to behold.

The person who delights in Paris might turn up his nose at smelly Rome or odorless Copenhagen. An aficionado of deserts might prefer the Desert of Maine to the shamelessly overbearing Sahara. For me, the cold waters of the North Atlantic evoked something deep and kindred — something, I daresay, that the waters of the vast and infinitely more ancient Pacific did not. Glacially scoured boulders put my feet in an inspirational mood; forests and grasslands did not. Resolutely barren islands made my soul sing; the more barren, the more rollicking the song. Why, I wondered, did I have such an affection for high-latitude places? Simply because the thin air and strong winds seemed to enhance me? Because the brute carcinogenic sun was less in evidence? Because northern people might actually be composed of at least three parts brilliant adamantine rock? I didn't know. So I decided to investigate this curious aspect of myself by undertaking a much longer journey, a journey that would carry me across the entire breadth of the North Atlantic, from the Old World to the New, from Norway to Newfoundland.

Norway to Newfoundland! This route, I quickly realized, was a subject in itself. For it was the route the Vikings took when they set their clinker-built boats in the sea. I resolved to use the wanderings of these cranky, restless people as a rough travel itinerary, trekking the same gaunt cliffs they trekked, camping on the same wind-scudded skerries they camped on, breathing the same salubrious air. But I would play the occasional game of traveler's roulette, too. Since I couldn't very well ship aboard the same boats as Eric the Red and his companions, I would ship aboard any boat I could find, even if it sailed somewhat off course. Also, between a fast boat and a slow boat, I would choose the slow boat. Between a slow boat and walking, I would choose the latter, as walking makes the world the vast and savory place it used to be in times long past.

To follow the Vikings meant following the sun in the direction of White Man's history. Like the multitudes of emigrants after them, they left old corrupt Europe for lands of milk and honey. Only difference was, their notion of milk and honey happened to be geological — ridges of sullen infertility rather than gentle plains, fang-shaped islands rather than cities, and ontological rock. Rock that offered them new worlds even before it offered them the New World. For it was only during the interglacial warming period of the late Quaternary, a trifling 10,000 to 17,000 years ago, that the North Atlantic began to emerge from beneath its huge carapace of ice. With just a few exceptions — seabirds, seals, some coastal plants — every living thing from Canada to Greenland, Iceland, and northern Britain had been killed off or sent south by the ice. Only after it retreated could mammalians (including the Vikings) travel to the lands it left behind, lands ribbed with moraines and littered with till. Lands where the milk was water from glacial tarns and the honey was the scent of bird-fretted sea cliffs.

Following the Vikings also meant traveling from one wide open space to another:

In the ninth century the fjord country of western Norway suffered from a common malady — too many people and too little land. Also it suffered from the attentions of an autocratic monarch, King Harald Finehair (862–930), who had a penchant for high taxes and centralized government — a penchant that finds its sequel in the latter-day Scandinavian welfare state. Norwegians had to put up with being hemmed in both politically and physically. After King Harald won a notable victory at the sea-battle of Hafrsfjord, a number of them decided to turn the prows of their longships *vestan um haf,* west over the sea, toward the setting sun.

The first group of wayfarers sailed north of Britain, where they settled in the Orkney, Shetland, and Faeroe Islands. Soon these island groups (smaller in area, taken together, than the state of Connecticut) began to seem a little crowded, for in those capacious days two or three thousand people was considered a thunderous horde. Many of the wayfarers moved on, and Iceland became settled. *Too* settled: by the year 950 the island's few decent roods of pasturage had been snatched up. In 982 Eric the Red sailed two hundred miles west from Iceland and sighted Greenland. Later he came back bearing tales of a green and fertile land (justifiable license, perhaps, with Iceland in the throes of a famine), uninhabited (he was dead wrong about this), and just waiting for colonization. And colonists came by the longboat-load. Soon Eric's settlement was itself too populous, and a new settlement took root a few hundred miles farther up the coast, near present-day Nuuq. Meanwhile Eric's son Leif set out to prowl the shorelines of Labrador and Newfoundland for still newer lands, newer sanctuaries.

Space, autonomy, elbowroom. In their never-ending pursuit of these ideals, the Vikings possibly journeyed as far south as Memorial Drive in Cambridge, Massachusetts, where a plaque commemorates an alleged visit by Leif Ericson. They may have made it even farther south: I know of one geographer savant who insists that the Vinland of the Vikings was just a few miles north of Fort Lauderdale, Florida. Doubtless Leif and his people *could* have visited either place or settled anywhere else they had an eye for, including the Côte d'Azur, which they knew about from their grandfathers, who had burned down the French Riviera. Instead they chose places that were hypaethral, unkempt, forbidding, or just plain empty. They chose the last places at the very rim of the globe, past which the sea suddenly turned into a downward torrent; but rather than falter, or tumble downward themselves, they simply sailed on and discovered new last places. Their journeys were a celebration of the North Atlantic wilderness. My journey would be, too . . .

And so it was that I gathered together my gear, stuffed a few changes of clothing into my rucksack, and flew from Boston to Bergen, Norway. The day after I arrived in Bergen, I climbed up the gangplank of the twenty-knot ocean-going steamer MS *Norröna,* of Faeroese registry, bound for the Shetland Islands and points north. It was a beautiful spring day, rainy and stone-gray, just the sort of venerable North Atlantic day that makes a person feel good to be alive.

Chapter 2

BIRDS OF A DIFFERENT FEATHER

TEN HOURS OUT of Bergen the *Norröna* came in sight of the Shetland Islands.

The Shetland mainland is a long attenuated backbone of peat and stone shaped like a dagger and pointed toward the British mainland. The offshore islands lie helter-skelter in the sea like a bunch of mercenaries undecided whether to stab the British lion or float off to join some other hireling cause. One of these islands, Foula, sits by itself, impervious to such nonsense, a chunky meditative bulk brooding only on its own diehard rhythms. Indeed, Foula sits so far off on the horizon that it was once thought to be the legendary Thule. It wasn't, and isn't. The only authentic Thule is the giant American air base at Cape York, Greenland, whose radar commonly transposes flocks of migrating geese into a Soviet nuclear attack.

Foula has always been a little too far away. It was overlooked in 1469 when Norway pawned Shetland and Orkney to Scotland as part of a royal dowry, a slip that allowed Foula to continue on, with its own Norse king and queen, until the middle of the seventeenth century. It became British largely by geographical default, since Norway hardly knew it still possessed a colony in the British Isles. Being British (or Scottish) did not mean a thing to this pawky little piece of Old Red Sandstone so hermetically sealed off from the rest of the world that it was omitted from a good many maps. Foula was not on the road to anywhere. Even now Foula remains the odd island out in a group of odd islands out, the Shetlands, which belong to Scotland, itself on the fringe of the United Kingdom. North Sea oil has brought Shetland more or less to the public's attention. Not Foula: it is still left off the map occasionally.

I decided to inaugurate my Viking trip on Foula because it was the last place in the British Isles where Norn, the old language of the Vikings, was spoken. In 1759 an Englishman named Low visited the island and took from the lips of one William Henry of Guttorm a thirty-five-stanza Norn ballad about the daughter of a medieval king of Norway. The nineteenth century saw the demise of Norn as a living tongue, yet I'd heard rumors on an earlier trip to Shetland that odd snatches of the language were still being spoken in thick-walled houses by the squally sea, among a company of secretive old men. If true, this would be a remarkable thing in the Britain of Liverpool Eight, Paki bashing, and punkhood, rather like finding a passenger pigeon in a contemporary zoo.

"Foula?" said the skipper of the *Norröna* over a glass of schnapps. "Haven't the lights gone out on Foula?"

"Not quite," I said.

Here was a man who'd sailed as far south as the Roaring Forties and as far north as Sisimiut in West Greenland. That he knew so little about this particular outcropping in the sea gave me, I must admit, a certain pleasure and confirmed (if I needed confirmation) my decision to visit the island. Later we checked his charts on the bridge and there was Foula — graphically not forgotten by at least one cartographer. All around the island WKS (wrecks), RKS (rocks), LDGS (ledges), and OBS (obstructions) warned the prospective mariner to stay away. Foula's waters were a dazzling minefield, which made the island inhospitable to yachtsmen, day-trippers, and even Her Majesty's public works brigade, though not — I learned in a few days — Jehovah's Witnesses.

At two A.M. I got off in Lerwick, Shetland's tiny capital, an enclave of gray granite rowhouses and staunch churches peering from their hillside roost down on the prickly antennae of the fishing fleet. Like all good towns at this hour, Lerwick was fast asleep, but like all sailor's towns, it was also wide awake. Little roving bands of Russians and Norwegians had received shore leave from their factory ships and were looking for a good time. One of the Russians stopped me and asked somewhat forlornly where all the girls were. "At home in bed," I replied. He looked even more forlorn. Clearly he felt a sailor's town ought to have at least one or two waterfront whores to lend it some respectability.

The Norwegians, on the other hand, had been coming here for too long to expect more than booze from a Shetland waterfront. A thousand years ago the islands offered them refuge from King Harald; more recently, in the 1940s, the so-called Shetland Bus ferried them across the North Sea after the Nazi invasion of Norway. Now their visits were less politically charged, unless you consider the depletion of fish from Shetland waters a political issue. Lerwick must have made them feel right at home, with its Norwegian lifeboat, Norwegian welfare center, and fishmeal factory run on Norwegian capital. Local shopkeepers cheerfully accept this capital — much more cheerfully than they accept Mrs. Thatcher's debased pounds.

Since there wasn't any action on the waterfront, apart from a few teenagers trying to neck and eat greasy chips at the same time, I took to the road. In lieu of the morning bus, I decided I'd walk the fourteen miles to Walls, port of call for the Foula mailboat. For I wanted to renew my acquaintance with the bald glaciated Shetland earth, remind it of my presence, let it sniff the leather of my boots and feel their stomp, so that when it came time for me to visit Foula's steepdown cliffs, this earth would recognize me and not be inclined to fling me overboard twelve hundred feet into the sea. A bootless prisoner of a motor vehicle might not be so lucky; he might be perceived as an undesirable alien and, once he left his cage of metal, be catapulted to his doom.

Shouldering my rucksack, I climbed up cobbled lanes so steep they required handrails and threaded a maze of dark wynds and rowhouses. Near the Iron Age broch of Clickhimin, which stood across the road from a modern housing estate, I paused to chat with a man walking his dog. He was a baker, he said, and no more terrible ("*Turr-r-rible,* laddie!") profession existed in Shetland. It was well-nigh impossible for him to get apprentices anymore. All the young lads were heading up to the Sullom Voe oil terminal, where they earned three times as much as they could earn in his shop. But wasn't a good loaf of bread, he asked me, more valuable than a big blob of North Sea oil? He opened his shop and gave me a loaf of soda bread with the specific gravity of lead. Then we went our separate ways, I overland to Walls, he — it would appear — to an early retirement.

I'd met quite a few other Shetlanders who admitted to a powerful dislike of the side effects of oil. They muttered in their pints about the widening of roads, dead seabirds, crowded pubs, crowded shops, ugly housing estates, overpriced wares, and the accents of imported British and American oil workers. *Accents!* When local folk fuss about someone's accent, you know there's more to it than meets the ear. You know they're not just fussing because the Scots part of them likes to fuss. Rather, they're worried about an assault no less deadly than an assault by a lockstep army: an assault to the gut of their own pridefully rundown and outmoded way of life. Perhaps they're also trying to drum up a little support for a species —

themselves — threatened by extinction. What oil supplies to the community chest, it takes from the marrow.

Yet the farther you travel from Lerwick, the less evidence there is of this upstart oil. The only accents tend to be Shetland accents, triumphantly impenetrable. And except for the main north–south road to Sullom Voe, the roads are all narrow twisting lanes on whose asphalt a thunderous oil lorry would be an impertinence. Also, it wouldn't fit.

After I left this main road, I climbed up and down a series of heather-and-sedge hills bulldozed clean by Shetland's own miniature ice cap during the last Ice Age. I cut across a switchback and scared off a few sheep who were busily cropping the grass down to its roots, a reminder of why there are no trees, scarcely even bushes or shrubs, in Shetland. Soon the wind (another reminder why there are no trees) picked up and my cap was whisked into a small loch. Five minutes later it was blown off again, now into a peat cutting. Fifteen minutes later it was impossible to light my pipe even with my supposedly windproof matches. In half an hour my ears felt like they'd been tacked to the amplifier at a rock music concert. Yet this wind was a gentle breeze compared to the cyclonic blasts that sweep eastward from Iceland and strafe Shetland with their artillery. In 1963, 177 knots were registered at Saxavord on the isle of Unst before the wind seized the anemometer and carried it off to parts unknown. I had heard accounts of brawny men thrown off their feet and babies wrenched from their mother's arms. And on my earlier visit to Shetland, when I'd gone to the island of Fetlar to observe snowy owls, I had heard this story:

In the early 1970s the roof of a house belonging to an elderly Fetlar woman was blown away. As if that weren't bad enough, the wind whirled down and picked up some of her possessions, including a cache of letters, and distributed them about the island. The woman had been known as a person of lifelong ascetic, if not resolutely virginal, habits. The letters, written years before, proved otherwise. These ardent declarations of love to a certain sea captain contained detailed notes showing that their union was not strictly platonic. Everyone on the island possessed elaborate testimony to this fact. They mentioned it, not disapprovingly, to the old woman, who by slow turns died of shame (her lover had died many years earlier, in the Great War).

After Tingwall, the site of the old Viking parliament, the road wound around arthritic inlets and perched atop gray sandstone cliffs, angling through one dark bog after another. Bog cotton, like spent spume, fluttered along the pavement. I saw a few of the celebrated Shetland ponies, their foals flung on the ground like rag puppets. They had the same off-white, skewbald, fawn, and brown colors as the sheep, and they were losing their coats in the same early summer fashion as those sheep, too. Shreds of hair stretched from their hocks or floated behind them in the wind.

Dawn was approaching with typical high-latitude light, a kind of perpetual muffled gray with a few brighter spots thrown in for variety's sake. At seven A.M., the first car I'd seen in hours drove up and stopped. A reddish head with charcoal eyes peered out at me. Obviously a head built for bulk so that the infamous Shetland winds would not dislodge it from its body and send it rolling over hill and moor.

"Hop in, laddie."

I was too close to my destination to hop in. Would Mallory, Irvine, or even Hillary have hopped in at this point? But the man would not be put off so easily. His massive head hung there with a coaxing grin on its face. I wondered if Shetland hospitality, a bequest from Viking times, *required* that he give me a lift, else the god Thor would exact a cruel and unusual punishment from him for forsaking a lone traveler on the road.

But then I heard him snoring lustily and noticed that his breath was 86 proof — another Viking bequest.

At Bridge of Walls I passed a number of derelict, roofless crofts which, if this were the tropics, would have been buried in the all-encompassing underbrush or at least strangled by a few creepers. Here they were *memento mori* that gave the lie to their own transience, not decaying naturally with the years but just *staying* decayed while less stolid organic matter was sucked back into the earth. Then I arrived in Walls itself, a tiny village that had been languishing ever since the salt-fish trade died. Walls was 1912 preserved in amber. Its streets had a quality of eternal peace denied to places carrying on through gainful endeavor. I saw only one very old woman; she was herding a very old cow. When I asked her where everyone was, she replied, "Everyone's drowned at sea or repairing helicopters at Sullom Voe . . ."

At the pier I sat down to wait for the *Westering Homewards,* the Foula mailboat. It came right on time, two and a half hours late.

Foula is twenty miles from Walls, but the sea makes it seem at least twice that far. A strong tide runs between the island and the mainland and has the odd habit, not of flowing in one direction and ebbing in the opposite, but of flowing from every point on the compass in turn. Thus the trip had the tumult of a tea leaf in a boiling pot. It was made even worse by the boat itself, a thirty-five-foot former Royal Navy lifeboat that appeared to have seen action during the Battle of Trafalgar. One of the two-member crew had to nurse its single-cycle Perkins diesel engine the whole way to prevent it from dying. The other member of the crew was steering by compass and solemnly informed me that radar was a "superstition." My only fellow passenger was an islander named Tom whose abscessed tooth had been fixed in Lerwick (nothing so extravagant as a dentist lives on Foula). We screamed at each other to be heard above the engine, which kept exploding into noisy life and sending up smoke rings in magical profusion.

Tom called himself an "Antipodean dropout," since he had come originally from Wellington, New Zealand. As a boy, he'd read *Treasure Island* sixteen times; this seemed to have had a formative effect on his later years. Good books, he said, were even more insidious than bad parents. He was trained as an epidemiologist — his people had been in the medical profession all the way back to the Maori Wars — and did his internship among the Fore tribe in the highlands of New Guinea. The Fore were then being decimated by a rare, extraordinary disease called kuru, transmitted through ritual cannibalism; pregnant women and children would eat the brains of their dearly departed to acquire their virtues, along with the lethal kuru virus. When the Fore were told to stop eating brains, the disease was cured. Tom's interest in a medical career was cured, too. ("In one year, 1971, Albert Schweitzer changed into Long John Silver.") In Port Moresby he married an Australian nurse and convinced her to homestead with him on the most isolated island he could find in his vest-pocket atlas. This turned out to be Pitcairn Island, home of the *Bounty* mutineers. They settled there but could never seem to get along with the Pitcairners, all Seventh Day Adventists who eschewed the dark satanic brews of tea and coffee for a more godly drink made from bran husks. Tom and his wife left Pitcairn, traveled for a while in her ancestral Scotland, and — lo! — ended up on Foula.

Every island imposes on its citizens a kind of tyranny wherein nothing is permitted to be on time, there's no food other than sardines in the local shop, and the one local doctor may not know the difference between open-heart surgery and the removal of a wart. Either you adapt to the oft-perverse demands of insularity, go crazy, or go home. As a result islanders tend to have thicker skins and more patience than the rest of us. They live at once externally and internally, with the rhythms of the seasons and the plushness and solitude of their own thoughts. Often their dance is awkward and a little stupefying, but at least it has its own steps.

Eleven years ago I was dropped off on Mingulay, a hilly rock of an island in the Outer Hebrides, by an old lobster fisherman named Hector. Mingulay had been wholly devoid of people since 1908, and its formerly busy pathways had become a complicated network of rabbit burrows. Eighteen miles from the nearest hearth, the island was a rapturous place if you knew you were going to get away from it, less rapturous if you felt you couldn't. Hector did not return as promised that evening, nor the next evening, nor the next, either. Six days went by before his nephew arrived with the scent of hot grease coughed up by the pistons of the old man's lobster boat. Hector had had a stroke ascending the steep slipway in Castlebay, Barra, and only the previous day had he regained consciousness in the hospital. His first stricken words were, "I've left a man on Mingulay . . ."

The first day after Hector didn't come, I felt panicky. I had brought along only two Cadbury's fruit-and-nut bars; I figured I would ration them to a bird-bite every few days and then throw in the towel. My cluster of bones would be picked bare by the jackdaws, who would appreciate the change of diet from nibbling on the bones of Mingulay's near-feral sheep. Only gradually did I realize that starvation wasn't the obvious solution to my problems. I began to forage for food and soon found I could gather sorrel and wild celery from the nettled hillsides and mussels from the rocks at low tide. I cooked up a limpet stew in an old rusted-out skillet scavenged from a seventy-five-year-old midden heap. I even killed a puffin, though somehow I managed to overcook half of it and undercook the other half, a difficult procedure with such a small bird.

It rained intermittently during the entire time I was on Mingulay. The only refuge was a decrepit sheep fank which sheltered several dead ewes in attitudes of writhing. After a while the rank cryptlike odor didn't matter so much (I can still smell it, though). My personal fate — just a few days earlier a blend of self-pity, fear, and anger at being left to rot on a desert island — didn't seem to matter so much, either. All that mattered were classic functions like eating, sleeping, and rising to wakefulness the next day. Each new morning pulled at me like gravity, and each night I congratulated myself on getting through the day. When the boat at last chugged to my rescue, I was relieved and exultant, but also a little confused. My pulse beat with the cadences of survival, not rescue. Mingulay had torn off a little part of me, and now she wanted the rest.

In that very short span of time, I think I achieved a habit of mind not unlike an islander's.

The approach to Foula was forbidding (rocks, ledges, reefs), and I could see for myself that one of the island's tyrannies was the absence of a good harbor, even the absence of a bad harbor. There was only a narrow geo called Ham Voe, through which tides rise precipitously. Ham Voe dries up at low tide, so *Westering Homewards* always has to be hoisted onto davits and swung onto a cradle, which is then hauled up with a hand winch. Foula's survival depends on this tedious maneuver, and on the muscles of pensioners cranking at the winch.

Evacuation has been suggested. Both winch and islanders appear to be old and rusty, and Foula itself doesn't seem to fit into anybody's future plans. But the notion of evacuation has been the albatross around the island's neck ever since a 1936 film entitled *The Edge of the World* was shot here. Directed by Michael Powell, the film was a mawkish love story about the abandonment of a lonely island off the north coast of Britain. The British public took it as a mawkish documentary and concluded that human life on Foula was, or shortly would be, extinct. Foula folk blame the film for many of their ills, including the fact that the Royal Exchequer doesn't seem to think they exist.

Yet they do exist, though a passing shower kept their attendance at the pier down to a straggle of old men dressed in dark frieze. Nothing much else to do on a day like this except stare at the mailboat, and they were doing that with visible wonder, as if astonished that it hadn't sunk. After we slammed into the pier, I hopped out and asked one of these men where I could pitch my tent without incurring somebody's proprietary wrath. Between complaints about the weather, the fishing, and the English, he instructed me to set up my camp in South Ness at a place called Guttorm. This seemed quite appropriate, since Guttorm was the site where the first Viking settler on Foula — a man called, not surprisingly, Guttorm — had set up *his* camp in King Harald's day.

So I hoisted my pack and started walking again. Almost immediately I began to observe ruined crofts of the sort I'd seen on the mainland, some merely roofless, others reverting to their stony essences. Many of these ruins were inhabited by fulmars, but others would not have provided a decent billeting to the Thule fieldmouse, a subspecies found only on Foula. The ruins were just heaps of rude stones, kept around because there wasn't much else to do with them, perhaps also because carting them away would be like carting away memories themselves.

All at once I heard a swooping sound and saw a blur of dull cinnamon just before a set of strong pinions whacked my head. I was under attack by another of Foula's tyrannies — a pair of great skuas (*Catharacta skua*), known locally as "bonxies." A bird with a five-foot wingspread and very nasty manners, the skua dive-bombs invaders of its self-proclaimed turf with the ferocity of a World War I flying ace reincarnated in feathers. As it dive-bombs, it utters its war cry: *Scar-r-r-re! Scar-r-r-re! Scar-r-r-re!* In more tender moments, it may squat on its nest and sing: *Uk! Uk! Uk! Uk!* There is no sound in all the avian kingdom more unlovely than a skua's affectionate *Uks!* to its mate.

A theologian once asked the English geneticist J. B. S. Haldane what inferences he could draw from his lifelong study of creation as to the mentality of its Creator. "He has an inordinate fondness for beetles," Haldane said. Had he been asked this question on Foula, he might have said something about the Creator's fondness for violent seabirds. Not only does the island boast the world's largest colony of great skuas — upwards of 8,500, or 225 for every man, woman, and child on the island — but it also plays host to a colony of arctic skuas (*Stercorarious parasiticus*), a slightly smaller, dimorphic bird even more bloody-minded than its big cousin. The arctic skua loves to flex its talons, and it uses its wings like an airborne pugilist. A pair of these birds probably wouldn't be able to kill a full-grown man in normal health, but they might send him weeping and wailing back to his mother's apron.

I walked on. The assaults continued. It dawned on me that we may be peering at skuas through the wrong end of the telescope, since they bear a much closer resemblance to their pterodactyl ancestors than to, say, the gentle chickadee. They would starve rather than pick at seeds and grains, and they show a marked preference not for the neatly manicured suburban yard but for the most forbidding, inimical, lonely regions of the earth. The skua is the only bird that chooses to live in both the Arctic and the Antarctic, and it's the only living creature other than crazed explorers ever seen in the vicinity of the South Pole. The toughness of its habitat emerges in the toughness of its behavior.

Fortunately Guttorm was perched by the sea, and skuas don't

much care for the direct bite of salt air. I pitched my tent and trenched it, a necessity in these up-and-down, liquidy parts. Once in northern Iceland I neglected to trench my tent and awoke to find my air mattress afloat in a not inconsiderable body of water. My down sleeping bag was waterlogged and therefore useless. The writing in my notebook had blurred into scribblings from an ancient alphabet. My food was a broth, and my matches, which I'd put in an empty center-fire cartridge case, had floated away, perhaps downstream to the Arctic Ocean. I was, to put it mildly, miserable.

Tom, my friend from the mailboat, had me over to supper that night. He and his wife Flossie dwelt in a thick-walled stone house set off by itself, like almost every house in Foula, claiming the inalienable privacy of an island where no sensible person would want to live. Near the half-door, lambs, cats, chickens, and children were in the process of getting themselves so gleefully muddy that they all looked like residents of a Carboniferous bog.

"First-rate walls," I said, fingering the thick stone.

"They're not thick enough," said Flossie.

"Not thick enough? For what? Repelling a warhead?"

"No, repelling the flanns. They come right through the wall and rattle our dishes."

These flanns occur when Atlantic gales hit the island's cliffs and then are deflected over them, reaching near-tornado status by the time they hit the crofts below. Haycocks, roofs, and chickens are flung into the sea as if they had been picked up by an aerial suction pump. Legend has it that all the crofts on Foula were once situated in a village, but an unusually powerful flann came along and scattered them to their present locations.

A neighbor named Bobby joined us for the meal, saying that cold mutton was his favorite food. Bobby was a lantern-jawed man of seventy or so whose conversation mingled native superstition with incisive jabs at royalty and heads of state. He mentioned an ancestor from the isle of Yell who'd been burned at the stake because he caught his fish already cooked. Maggie Thatcher had the crooked grocer's grin of the Mona Lisa. Nothing that old suckbutt Heath touched ever blossomed. Then there was the Foula man who fiddled for a wedding of trolls and crumbled to dust the next day when he received the Holy Sacrament. Bloody fool Charles I got what he deserved, hiring a ventriloquist to imitate the sound of a sheep being slaughtered. The daftie! And if these are such advanced times, laddie, then tell me this — Why can't we catch *our* fish already cooked?

Bobby had the laugh of a spry old hyena. He, Tom, and Flossie seemed more or less content to inhabit their insular fingertips. It was true that fumes from Tom's tractor gave Bobby diarrhea and Bobby had once slit the throats of two of Tom's sheep by mistake and *Westering Homewards* sometimes missed the island completely in the fog and went sputtering off in the direction of Greenland, but this was Foula, man. What do you expect from the edge of the world? Tom's kids had 4,000 acres of hill and bog to call their playground. Small matter that the schoolhouse had been condemned in 1956 but was still in use. Small matter that in oil-abounding Shetland, Foula had perhaps the highest fuel rates in the United Kingdom, for the island boasted enough peat reserves to last until the Day of Judgment. And a room filled with peat smoke was thought by the old people to be almost too healthy. They could never get enough of it in their eyes, hair, and clothing, even attributing their longevity to peat smoke.

After the meal we adjourned to the sitting room (a couple of chickens were sitting there, too), and I inquired about speakers of Norn. Tom mentioned one or two elderly men but said I'd have to bide my time, since most of the old people were extremely wary of visitors. They thought the outside world one enormous breeding ground for plagues and epidemics, against which they had almost no resistance. This was because the so-called muckle fever, smallpox, had once killed off all but six islanders. I'd have to be observed through squinting windows as a fit and sturdy person, suffering from neither whooping cough nor the common cold, before they'd consent to speak with me.

"But the muckle fever was in the 1720s," I said. A minute later I realized that Bobby had been keeping the distance between him and me to a minimum of six feet.

June 1. Bright morning, as luminous as a pauper's last penny. A soft haze clings to the conical, heath-clad hills, barren only in their absence of trees. To be camped among these hills is to be camped in Eden minus all the damned creeper vines.

After breakfast I stroll off for a day or two of scrambling on the hills. My walking stick I carry not in the customary manner but well above my head, as the skuas always think the highest point is the head, no matter how skinny and unconventional its shape. Soon a pair of arctic skuas notice me clumping through their nesting ground. In time-honored fashion they start screaming obscenities, and after they've caught my attention they begin injury feigning with a show of stricken wings. (Foula folk say they're not injury feigning at all, but registering shock that a lesser mortal would dare invade their turf.) Once I'm at a safe distance from their eggs, they change tactics and proceed, one after the other, to dive-bomb me, attacking not from the front, where I can see them, but from behind. Each time I turn my head they change their angle of attack to the stern. I cover my nose; I once met a man in Iceland who had had the tip of his nose taken off by one of these birds.

Higher ground and I'm in bonxie country, each pair standing with its clutch of olive-green eggs on its own tussock mound, mounds nitrated into fertility by years and years of guano. A pair spot me and waddle off to launch themselves, zooming into the air with the buoyant lift of fighter planes. Once aloft, they adopt the same tactics as peregrine falcons, circling in opposite directions to get above me, even turning away before they drop down to dive-bomb me simultaneously. I get a wing across my forehead and a shower of shit down my neck. Every fifty or so paces a new pair take up the attack and the old pair return to their mound, raising their wings like a greeting of centurions.

The whole skua scene makes Alfred Hitchcock's *The Birds* seem as precious as a Jane Austen tea party, yet I alternately climb and stumble on because, well, that's what you do on Foula. I move along the gaunt heathery spine of Hamnafield and pass a group of *planticrubs,* circular stone structures a bit like Irish beehive huts, though these were built to store cabbage and kale, not monks. Like the rest of the island, the *planticrubs* strike an ideal kind of balance, half in ruins, half still functional, so that you're reminded of life and death, cabbage soup and mortality, together. I walk by the stony remnant of an old Viking well, now dry but once noted for its healing powers. Then I pause at the cairn marking the so-called Hole of Liorafield, which, according to rumor, cuts through the center of the earth. A story is told about a dog running after a sheep; the sheep jumped into the Hole, followed by the dog, and they both came out in Madagascar.

Now two bonxies race past me in hot pursuit of a kittiwake, rising and sinking like twin evil geniuses out to corrupt a soul. All at once the kittiwake disgorges a fish, and one of the bonxies swoops down and makes a perfect midair capture. Skuas have too much piratical pride to get down in the water and fish for themselves. They steal from other birds (hence the label "kleptoparasite") and often dine on other birds, too. In Faeroe I once saw a skua extract a puffin from its burrow and fly the hapless bird down to sea to drown it. This drowning took quite a while: the Law of the Jungle has a palate no less demanding than a patron at Maxim's.

I'm climbing another tyranny now: The Sneug. Twelve hundred feet below me breakers crash against the vertical cliff face with such ferocity that a bit of a spindrift fuzzes up over the top. I toss down a rock and it takes fully twelve seconds to reach the bottom. A man in his prime would take a little longer, since he'd be smashing into Old Red Sandstone again and again before attaining his watery reward. Many Foula men in their prime did die in this fashion, and all for the sake of a bit of seabird in the family stewpot. They'd lower themselves down by horsehair rope, and maybe the rope would break or maybe it was already frayed or maybe they'd even attempt a ropeless descent to impress the ladies. There was once a saying on Foula: " 'is granfaither guid over, 'is faither guid over, and 'e must expect to gee over Da Sneug, too." The rudest insult was telling a person: "*My* faither guid over Da Sneug, but *your* faither died like a dog in 'is bed." Thus with a few quick whisks of the tongue was your whole human birthright denied.

Bivouacked for the night on The Kame, highest or second highest sea cliff in Britain, depending on whether a Foula man or a St. Kilda man wields the measuring tape. I dine on somewhat whiffy Double Gloucester cheese, soda bread (courtesy of the Lerwick baker) hard enough to chip a bull rhino's tooth, and a bottle of Newcastle Brown Ale. Hard to imagine more princely fare or a more beguiling ambiance in even the most Michelin-starr'd of chophouses. Then I curl up for the night on a bed of singularly delicate blue cuckoo flowers, none bigger than a baby's toenail.

June 2. Another lovely day, with a feathering of light air on the sea. Bread and sardines for breakfast. The bread is now hard enough to chip off a mastodon's molar. What can I say equally laudatory about the sardines? Only that certain well-bred Londoners once believed there were vintage years for sardines, as for wines, and that Oscar Wilde's son Vyvyan Holland founded a vintage sardine club whose members owned cellars and held reverential tastings. It was determined that 1959 was easily the best postwar sardine year . . .

Down a steep, moist, heathery slope cysted with outcroppings of mica (Foula's lone industry in former years was the export of mica schist millstones to the Orkneys). At one point I stop for a rest and find myself being stared at by two gray-black Shetland sheep whose wool trails off them in a bewilderment of clots, knots, and gnarls. For which untidiness they should not be ashamed: Shetland sheep are the descendants of the peat sheep kept by Neolithic men and are the most primitive domesticated sheep in the world today. They don't seem to be ashamed, either. They stand before me unchanged, unflinching, even a bit disdainful, as if they have some sort of racial memory of my forebears hunkering around with loincloths over their manhoods. Or maybe they just regard me — scraggly beard, bread crumbs, birdshit, and all — as the most primitive domesticated human they've ever seen.

Back in the lowlands I leapfrog a field of tousle-headed tussocks composed of heaved-up sedge and new grass, which the French, ungallantly, call *têtes de femmes*. I reach a bog marginally more suitable for walking and shortly come upon an old woman cutting peats. Though she must be at least eighty years old, she has the physique of a bantam wrestler, and I'm certain she could bring me to heel in at most two falls. When she sees me, she draws her paisley scarf quickly over her nose and mouth (smallpox? leprosy? hiker's breath? the outside world?), but says, through the scarf, "Yaw, my dear. A fine day in da bog is a piece o' heaven."

Between the townlands of Soberlie and Harrier I wander through a whole metropolis of derelict crofts. Derelict of people, that is, for in every inglenook, cranny, and collapsed threshold sits a nesting fulmar. On Foula Einsteinian physics is translated into an aviary: nothing is lost that doesn't return tenanted by birds. The first fulmar arrived in 1878 on the back of a dead whale, and since then they've taken over nearly every rundown neighborhood on the island, often at the expense of the usually fearless skuas, in whom a young fulmar puts the fear of God. It looks rather harmless, just a soft billowy ball of fluff, but the young fulmar spits up a half-digested oily ooze that smells like rotting fish in its final, apocalyptic stages. The chick begins life by spitting at its parents, and soon it has learned to spit at everything, literally everything, except its parents. Try as you may, you can't wash this vile stuff from your clothes. But at least you can change your clothes. If you wear only wings, your flying days are over just as surely as when you get stuck in an oil slick.

Along the way to Guttorm I watch a raven and a bonxie fight over the afterbirth of a cow. The bonxie wins. I watch two arctic skuas chase a bonxie with upside-down, right-side-up pyrotechnics that put me in mind of Ringling Brothers acrobats. Afterward the two fly off squawking and chirruping to each other, as if to say: Next time the big fellow dies. Then I meet a kestrel-faced man on a tractor that looks like it's been hit by something nuclear. The tractor is pulling along a broken-down Morris Minor that looks even worse. The man tells me that he got the vehicle on the mainland, paid ten pounds to ferry it to the island, and is now getting ready to dump it in the sea, since the Shetland Islands Council will pay eleven pounds to anyone who disposes of a junked automobile. His net profit for this undertaking, he informs me not without a trace of pride, will be exactly one pound.

That's how they live on Foula.

June 3. Visit to the local blacksmith, Harry Isbister, who fashions items out of copper salvaged from the wreck of the *Titanic*'s sister ship, the RMS *Oceanic,* which foundered on a reef off Foula in 1914. A wreck of that magnitude was just about the kindest thing that could have happened to the island in those destitute days, Harry says. He would have given his wife and kids to witness it. I asked him about Guttorm. "Never met da man," he says, but then he goes on to add that at Guttorm's and my mutual campsite there had been a croft with a fire burning continuously in its hearth from Viking times right down to the early 1960s. Now that fire was extinguished, the croft was gone, the people were dust, and the land itself was bracken and nettles. *Ah weel,* he says, *it cannae be helpit.* And then he fries me up a rabbit.

June 4. Heavy rains and a shrieking gale. Last night the wind threatened to pick up my tent, poles and all, and deposit it in the sea. Had to guy it down with rocks the size of boulders. Lying awake I thought about the seamstress who had sewed the tent in some cramped Third World sweatshop. Maybe it had been Friday afternoon and she was eager to be getting home after a hard week and she hadn't bothered to backstitch that one critical seam. PRESTO! Face to face with an angry Shetland hurricane.

By midafternoon the winds have moderated from a Force 9 (strong gale) to a Force 6 (strong breeze — umbrellas opened with difficulty). Spent a quiet day at home reading a book on loan from Tom about Arab camel racing, of all things. Apparently the lightest and best camel jockeys are five-year-old boys purchased from their parents in the wilds of Baluchistan and then flown direct to the Arabian Peninsula. They lose their names and become identification numbers, which are sewn onto their shirts with strips of Velcro. Their saddles are constructed of Velcro, and they have Velcro on their trouser seats so they won't fall off. The most typical sound of a camel race is the sharp rasp of Velcro when a youthful rider dismounts his steed.

Dusk at midnight. A shooting star traces a parabola of soft blue across the suddenly clear sky.

After a few more days on Foula I became accustomed to the skuas. I even grew to appreciate them for the quickening they brought to my blood. I could take very little for granted with them disputing my right to their land. The sharpness of my focus increased; with the rush of anxiety, I learned to see as through a glass darkly the respective personalities of nearly every bird who dive-bombed me, some so proud of the broad athletic sweep of their wings that they seemed to attack with a narcissistic glee, others diving quickly as if they just wanted to get it over with, and even a few whose attacks were so poorly synchronized that they swooped down on a collision course with each other.

On one of my walks I saw a man near Mill Loch cutting peats with a *tushkar,* a double-edged spade that allows its user to make a cut in the bog and wrench up a trapezoid of peat all in the same easy motion. I watched him dig with this *tushkar* even as he was being dive-bombed by a relay team of bonxies. He seemed to have timed his actions perfectly; without bothering to look up, he ducked in accordance with each bonxie's dive, as if man and bird were connected by an invisible string. Maybe after thirty years on Foula I'd develop the same exquisite timing, and the same string.

It was this man who told me that the word "bonxie" in Scots meant "fat old woman," referring to the manner of these big ungainly birds waddling around their tussocks. This seemed to me a peculiarly generous way to regard such a bloody-minded creature, but you have to be generous in a place like Foula, else all its tyrannies would take you to task.

Meanwhile Tom had been making a few inquiries about speakers of Norn. Nearly everyone he talked to seemed to agree that I should visit a man named Andrew Isbister. In his seventy-odd years, Andrew had left Foula only once and that was just before World War II, when he'd sailed around the world on a tramp steamer; he'd rummaged around places like the Seychelles and the Maldives, but he'd never once set foot in England. A crane-lift accident had mangled his right leg and sent him limping back to the island where he tended his sheep and, until recently, fetched paraffin for the lighthouse. In 1979 he had recited a burial formula, the Lord's Prayer, and a poem about eagles to the German philologist Reinhold Weber, who was following up turn-of-the-century research by Jakob Jakobson, a Faeroese vocabularian, to determine how much Norn survived in Shetland. Like the last native speaker of Manx, who reputedly spoke that now-defunct Celtic language only to his parrot, Andrew spoke Norn to his sheep just to keep his tongue limber. Or so the story went. The story went on to suggest that the sheep baa'd Norn back to him in their own inimitable accent.

I visited Andrew's *butt-an-ben* in Mucklegrind. Balanced on a knoll, the small whitewashed cottage seemed to be missing half of its slates as a result of last winter's storms (the other half looked as if they planned to be missing at the earliest opportunity). Inside I saw a rickety dining table with only a bottle of Stockholm tar resting on it. Did he perhaps eat the tar? It couldn't be much worse than the clay still eaten for its nutritional value in parts of the American South even today.

Finally Andrew appeared, still hobbling from his crane-lift accident. He had a face of wind-devoured granite, and his white hair stood out from his head like lengths of thirteen-amp fuse wire. He wore an ancient frieze coat that would have made an excellent nest for a family of rats. Squinting at me, he said, "Weel, man. Are ye a Jevohah's Witness, then?"

It seemed that Jehovah's Witnesses occasionally sailed over with the *Westering Homewards*. Just as they ransacked big city slums and the Third World for converts, so too they proselytized for their faith on remote Foula. But they seldom had much luck, and the skuas usually conspired to drive them out before they could wreak a conversation on one of the locals. Andrew showed me a copy of *The Watchtower* someone had dropped off on his doorstep a few months ago. (One week later I would see a copy of *Vagnóp* in the Faeroes and two months later a copy of *Vagttoramet* in Nuuq, Greenland).

All during our conversation Andrew kept himself several paces to my starboard. I stood near the big open inglenook and he stood near the door, as if getting ready to escape the pestilence I doubtless was bringing to him. Once, when I coughed, he gasped audibly.

"Yaw, yaw, laddie. A guid-willie-waucht o' Stockholm tar gies a fine smoky touch tae fish an' cabbage. Ye've never haed it? Weel, man. Whaar ha' ye been?"

When I told him I'd like to hear some Norn, he agreed to rattle off a bit for me, though I'd have to keep my distance, for I seemed to be nearly at death's door. Then he began to sing a song he'd picked up from two twin sisters, aged ninety-four, back in 1929. ("Bonnie wimmen dey was, even den.") His voice was like the clapperclaw of a crow. He was toothless, and his accent was almost impossible to penetrate. Nonetheless the performance had a curious, poignant, even otherwordly quality, suggesting not a human being so much as a stone on The Sneug suddenly giving itself over to song.

After he'd finished the last syllable, Andrew told me that the song was a lullaby. A mother is trying to calm her restless babe. She's cradling it in her arms and telling it to go to sleep, go to sleep, my sweet li'l bairnie, and if you don't go to sleep, I'll bang your head against the wall and *then* by God you'll go to sleep. A quite Viking sentiment, Andrew thought.

A little while later Andrew's sister dropped in and told us *The Earl of Zetland* was just then puffing into Ham Voe. *The Earl* was a cargo vessel that made special trips to the out islands, and today was a quite special trip because it was bringing in Foula's first mobile home ever. We headed down to Ham Voe, where the rest of the island was already gathered for this gala event. There's no cinema on Foula, but the off-loading of a mobile home would seem to be the next best thing — less star-studded, perhaps, but just as suspenseful as a good thriller. Will it go or won't it go? The winch wouldn't catch, the shiny blue-and-white oblong box seemed disinclined to leave the boat, and when it did, its journey along the logs placed on the pier for its benefit was at best haphazard. This mainstay of the British seacoast was supposed to be the new home for an islander recently evicted by his wife, but the procedure for getting it to its final resting place in Hametoun was so arduous I wondered whether the warring couple might not reconsider their marriage.

"Da puir bawdy," said Andrew's sister. " 'e'll be caald dis winter, A'm thinkin'. Da gro'll pierce dat thing a'maist lik a knife."

Then Andrew hobbled over to me and said: "All Foula folk ken Norn. Ax *him.*" He pointed to a freckle-faced little boy who might have made a very fine camel jockey and who at that moment was picking his nose with the concentration of a digger after buried treasure.

Andrew was right. In the end I realized that almost everyone on the island had snatches of the old Viking tongue in his possession. It survived in descriptions of what mattered most: animals, the weather, the sea. A *blaegit* sheep was white with black spots; a *katmoget* sheep had dark underparts; the poor quality of wool from a *katmoget* sheep was the *aliplukkens*; a *gaats* was a castrated bull (likewise a somewhat effeminate man). *Gro* was wind; *stoor* was a light breeze; a *gooster* was a strong breeze; *daggastö* was a wet wind; a *guzzel* was a dry, parching March wind from the southeast; a *pirr* was a light wind in July and August; *ungasstö* was a contrary wind; and the *flanns* came on like a tornado.

On and on it went, a whole vocabulary of hard, precise, visceral words: words that had been around for a thousand years and would perhaps be around for another thousand, owing to the strength and fortitude of their sounds. And if not, if Foula were to be evacuated or blown off the edge of the world by a particularly violent flann, there would always be the bonxie's *Scar-r-r-re! Scar-r-r-re! Scar-r-r-re!* which has a peculiar sort of strength of its own.

My last day on Foula Tom and I hiked up a thousand feet above Mucklegrind in search for some missing sheep. We paused at a knife-edged gap in the cliffs and he told me this story:

In the last century, a woman named Maggie had gone into service on the Shetland mainland, where she contracted leprosy. When she came back to Foula, her family refused to take her in, so she built a little stone hovel on the cliffs. Her father would climb up and leave bread and milk at her door so long as she promised to keep her leprous self hidden from him. Then one day he climbed up only to discover that the chunk of cliff with Maggie's hovel on it had dropped into the sea. If you listen hard, you can still hear the old man's cry of anguish upon learning of the loss of his daughter.

"A few weeks later," Tom said, "the old man went to the mainland with a boatload of herring. He caught a cold and in three or four days he was dead himself — which just goes to show that the mainland is, after all, a rather pestilential place."

Chapter 3

WHALE BASHING

IN LERWICK I picked up the *Norröna* again and sailed north to the Faeroe Islands. In one of the ship's three bars (that's three more bars than the Faeroes themselves have), I met an English filmmaker called Pinnock who was born on the river Ouse a few hundred yards upstream from where Virginia Woolf drowned herself. As a young boy he'd had nightmares of her body sloshing out of the river and politely joining his parents at tea. He'd even made a film about Virginia Woolf for the BBC to rid himself, he said, of her moist and bloated memory.

I offered to buy Pinnock a whisky. He declined, saying he'd just returned from the Land of the Fire Dragon, Bhutan, and his stomach still felt dreadful. Amoebas, he remarked, were the national food of Bhutan. For relief he'd scurried across the frontier into Nepal and a Katmandu waiter had asked him, "Apple pie or opium, sir?" and he'd made the mistake of taking the pie. The apples, he said, had been grown on gnarled old microbe trees specially cultivated in monastery gardens.

Here we were, in bouncy northern seas, discussing the dysentery regions of the globe. I mentioned the most uncomfortable time of my life, seated on a horse in the Andes and possessed by a malaise that opened the sluices at both ends. Pinnock said he'd once had a private audience with Somali President Mohammed Siad Barre, during the course of which he'd been attacked by a sudden and highly embarrassing stomach virus. I told him about an unpleasant experience I once had with a Mexican tossed salad, and he told me that in Bhutan tossed salads, not rats, carry the Black Death. Then, no doubt concerned about a repeat of his Bhutanese experience, he asked me what they ate in the Faeroe Islands. "Whale," I replied. "Whale, whale, and more whale. They call it *grind.* It's the only fruit that'll grow there." "Shit. And I left my stomach pump in Cheltenham, too."

Pinnock had been invited to the Faeroes as adviser to the newly constituted (1980) Faeroese TV station. For a fortnight his expertise ("I taught the Somalis everything they know about docudramas") would be at the mercy of the Faeroese TV commandoes. Mercy was my word, not his. When he queried it with a knitting of his nearly browless forehead, I told him the truth: Faeroese TV presents a view of the world such as Humpty Dumpty might have had if Humpty Dumpty had fallen down drunk on Saint Ólaf's Day.

You screw up your courage and switch on the tube. There's a newscaster discussing the American submarine that's been told it can't dock in Torshavn, the capital. The TV camera provides you with a close-up of the man's left ear. "*Já, já,*" he is saying, "the Løgting is following the example of New Zealand with respect to nuclear-capable American vessels." Now the camera freezes on the twin holes of his nostrils, where it remains for an unconscionably long time, before it moves on to the Honorable Atli Dam, prime minister, who says, "We are not afraid of the possible fallout with Denmark on this issue. We don't want to become a pawn in the East-West struggle." The camera assumes an angle that could have been invented only by the late Orson Welles during a moment of lofty inspiration. Shot from below, Hon. Mr. Dam's knees are considerably larger than the rest of his body, and his head is but a tiny pinprick. Switch to a sheep roundup. The newscaster is talking about a whale hunt in Klaksvik. A pod of 112 whales, he says. The camera focuses on a sheep's cud-chewing face. Meanwhile the newscaster is talking about the new fish plant in Fuglafjördur. The camera presents its viewers with a shot of the new public lavatory on the island of Svínoy. The newscaster mentions this lavatory even as you see the weather map, which shows a heavy cloud cover over all eighteen of the islands.

"Sounds positively surreal," said Pinnock.

"It is. For God's sake, don't try to change it."

Twice before I'd visited the Faeroe Islands, which lie halfway between Shetland and Iceland, rearing themselves up from the sea like an armada of tilted layer cakes. On the occasion of my first visit, I'd been forced to spend the night in the school on Fugloy, northernmost island in the archipelago, because the fog was so thick I couldn't locate the spot where I'd pitched my tent. The other time I stopped over en route to Iceland and somehow ended up at a silver wedding party in Klaksvik. Now a silver wedding party in Faeroe is scarcely less robust than a Hell's Angels bash in distant California. That night I joined in a chain dance and stomped to the accompaniment of ballads about prominent medieval personalities such as Olav Tryggvason, the Norwegian king who thoughtfully christianized the heathen Faeroese. On and on we stomped, until the first gray drizzle of dawn — dawn *two days later*. Afterward my feet hurt worse than they had when scrambling over the scree and porphyry of Faeroese cliffs. My teeth hurt, too — I'd jarred loose a filling with all that stomping.

On Fugloy I'd heard the tell-tale cry of *Grind!* — a cry calculated to raise the adrenalin level of every Faeroe Islander regardless of age, sex, or infirmity. The first Faeroese public election in 1850 had to be postponed because the cry of *Grind!* was raised just as polling began. I've been told about a Torshavn surgeon who heard someone yell *Grind!* and immediately dashed down to the harbor, leaving his patient cut open on the operating table. (Someone else told me the same story about the patient, cut open, dashing down to the harbor.) I dashed down to Fugloy harbor myself, but it was a false alarm. The *grind* turned out to be only some sharks swimming in the channel between Fugloy and Svínoy, so I wasn't able to witness a *grindadrap,* or whale hunt. This seemed a pity. For watching a *grindadrap* in the Faeroes is rather like seeing the Great Wall in China or the changing of the Guard at Buckingham Palace, though the spectacle is a bit more bloody.

On these earlier visits to the Faeroes, I'd been somewhat surprised by the riches I'd seen. A self-governing community within the Danish realm, Faeroe is perhaps the only colony in the whole sorrowful history of colonialism to be better heeled than the mother country. The lucky Faeroese have it both ways; they get public assistance from Denmark and a largesse of *kroner* from a seller's market for their fish, scooped from the sea by a high-tech (computers cut the bait) fleet of trawlers that ranges as far away from the islands as Spitsbergen, the west coast of Greenland, and Maritime Canada.

These riches are quite *nouveaux.* Less than two generations ago the fishing was primarily inshore, and the Faeroese maintained the sort of poverty level you associate with Sicilian or Albanian peasants, not northern Europeans. Even now they treat the spawn of these riches with a certain cavalier indifference, as if all automobiles, appliances, and techno-appurtenances were simply toys, easy to replace. A Faeroe Islander will go out and purchase a new Mercedes (despite a whopping 100 percent import tax) and drive it all around his stark Tertiary back yard, road or no road, as if it were designed for cattle herding rather then gracious living. To do sixty miles an hour in a terminal moraine with a Mercedes or BMW seems to be the goal of every young man; to impale that vehicle on a boulder the alternative goal. Such behavior is understandable, however. The first roads arrived in the islands only in the 1920s; it takes a couple of years to get used to them.

At six P.M. the *Norröna* put in at Torshavn, a town of 12,000 clawing at the basaltic slopes of southeastern Streymoy. A large assortment of Mercedes and BMWs was waiting at the dock, one of which hustled off Pinnock to do battle with the trolls of Faeroese television. Moments after I stepped ashore, I was embraced by a drunk whose breath, like Proust's madeleine, sent me back to times past. I could tell he'd been drinking the local pilsner laced with underarm deodorant, a mixed drink not uncommon in the islands, as the slight alcoholic tang of the latter gives a boost to the almost spiritless ingredients of the former.

The man was uncouth and unshaven, with a dead-planet look inhabiting his face. He embraced me again and called me his brother. I'd never seen him before in my life.

Poor fellow, I thought. He's caught between liquor rationing periods. Liquor is rationed four times a year. Only then and upon receipt of a form saying he's paid all his taxes can a Faeroe Islander buy his booze, which he proceeds to consume immediately, not gently sipping it whilst discussing theater, like the French, but *devouring* it by the quaff like the Russians. The rest of the time he'll drink anything in sight, including water.

The idea of rationing goes back to the last century, when the Faeroese showed a disturbing tendency to barter their land — bad acidy land, poor in minerals, it's true — for flagons of brandy. Nowadays the strongest advocates of rationing seem to be the austerely minded Plymouth Brethren and even more austerely minded Inner Mission Lutherans (I once shared a taxi with one of these latter austerities who, at the end of the ride, said: "Oh no. *God's* paying . . ."), but I've met some extremely faithful atheists who support it, too. Everyone seems to be afraid of skippers en masse swapping their trawlers for bottles of Haig & Haig or Stolichnaya.

Thus it is not surprising that Torshavn boasts none of the bars, pubs, watering holes, shebeens, groggeries, honky-tonks, caravanseries, nineteenth holes, or gin palaces that lend a certain rough civility to other towns its size. Nor does it have any other visible life after six P.M. unless you count the occasional Inner Mission boys' choir singing stern hymns to passersby on the streets. In the worldwide sweepstakes for most boring town, I'd place Torshavn directly below Guaranda, Ecuador, where the only available evening recreation is to head down to the Parque Montufar and watch the wasps crawling in and out of their nest in the armpit of a statue representing, I think, Liberty.

Not being in the mood for stern hymns, I caught the last boat for the green dromedary hump of Nólsoy, the small island that makes Torshavn a harbor by sheltering it from easterly gales. Once ashore, I made for the home of an old acquaintance, Marner Poulsen, a strapping jocular man who earned his bread from the sale of stuffed puffins. Marner was one of many Faeroese named for a prominent obstetrician, Marner Simensen, who specialized in births that kept both mother and child alive. In the old days, doctors seemed to think that if one of the two was salvaged, that was enough.

"Any *grindadraps* around here lately?" I asked Marner.

"*Nej, nej*. They've closed Torshavn harbor for hunting because they don't want any visitors to see us shedding the blood of whales. Torshavn gets one or two cruise ships a year now. It wouldn't be good."

Well, it wouldn't be all that bad. It wasn't as if the Faeroese shot whales Japanese-style with electronic harpoons and then processed them in factory ships. They weren't hunting noble leviathans like the blue whale, the bowhead, or that gay rutting troubadour of the deep, the great humpback. They only hunt the long-finned pilot whale (*Globicephala melaena*), which is not on the International Whaling Commission's endangered list and indeed seems no less healthy and thriving in Faeroese waters than cattle on a Texas range. Perhaps more healthy: at least *Globicephala* has the freedom to raft wherever it pleases and eat what it chooses before humankind draws the final curtain on its life.

Marner and I sat down to a supper of (what else?) *grind* and *spik,* whale meat and blubber.

Since my previous visit, Marner told me, a lot of foreigners had been coming around to provide the Faeroese with dietary instruction. They cited the brain size of the whale and argued that Marner and his countrymen should eat less intelligent cuisine. Not a few were members of various U.S. whale adoption projects for whom Flipper was the paradigm of all sentient life. They gave individual whales names like Patches or Ginger and wore cotton-polyester T-shirts that said "I LOVE MY WHALE"; they were rather sweet. But members of the Environmental Investigation Agency, a sort of sea-mammal FBI based in England, were not sweet at all. At a recent *grindadrap* in Vestmanna they had used gas-filled cylinders to force the whales back out to sea. And Paul Watson, the *Sea Shepherd* kamikaze, was in a class by himself. Recently he'd attempted to run his boat between the hunters and the whales; now he was trying to promote an international boycott of Faeroese fish products (McDonald's, for example, uses Faeroese fish in their fishwich). Rather than give in, the Faeroese had taken to wearing their own T-shirts, printed with this directive: VER FØROYINGUR — HVAL EN WATSON. (Be a Real Faeroese — Kill a Watson.) It was not likely they'd kill a Watson, though. Since 1877 they'd killed only two of their fellow human beings, and one of those was a mercy killing.

Many of these uninvited guests, Marner said, were highly vocal vegetarians. This did not surprise me in the least. Vegetarians have made a habit of crossing borders to campaign for their cuisine, as if the world were one instantly loamy field with the same softness of complexion as Ohio or the English Midlands. But less than 6 percent of Faeroe is arable. Having been cleared once by glaciers, it can't be cleared again. Most tender sproutlings would simply give up the ghost rather than struggle everlastingly with bad weather and worse soil.

"I *hate* carrots," Marner said, shaking his fist.

I watched him put the meat of his Viking ancestors into his mouth and wondered what evils might befall his digestive tract if he were obliged to change his diet. His big brawny physique reminded me a little of a giant panda, and the sad fate of the panda came to my mind. Though descended from carnivores and having a digestive system designed for meat, the poor panda eats only bamboo leaves, with the result that it leads a precarious life trying to get enough of these leaves to compensate for its physical inability to extract nutrients from them. I'd hate to see Marner and his countrymen stuck in the same culinary bind, or, like the panda, closing in on extinction.

"Would you keep up the *grindadrap* if McDonald's or Long John Silver stopped buying your fish?" I asked him.

"We would."

"But that would be a disaster, a return to the Dark Ages."

"You forget, my friend, that we spent a very long time in the Dark Ages. We must have found something of interest there."

On Marner's advice, I took the bus-ferry west to the island of Vágar and then climbed to Gásadalur, a lilliputian speck (pop. 25) on the map, the last roadless village in Faeroe. Even tiny Fugloy has a road, which it got a few years after my previous visit; one of the locals went out and bought himself a car, which inspired the busy government engineers to clear a stretch of nearly vertical hillside for that car to graze on. In the last decade sleek new tunnels have been blasted through previously immutable mountains to every remote little village, the idea being to keep the old traditional places alive and kicking. But the road in always becomes the road out, and nearly everyone seems to resettle in Torshavn, urged on, as one old ballad puts it, by "peppermint candy."

First the fog — on which everything in Faeroe is blamed, from incest to traffic accidents — came down and made it impossible for me to find the path, then the fog lifted and I realized there was no path to Gásadalur, only the occasional crumpled cairn. My hands splayed against the ground, I clambered up a series of steep, rock-strewn shelves called *hamrar,* which were moist from some sort of geological sweat. Loose stones and crunched-up basalt clattered under my feet like broken glass. Once I scrambled off in the wrong direction and almost plunged a thousand or so feet into a bare, lovely mountain valley. Another time I down-shifted to a three-point crouch and crawled unflinchingly — unlookingly, too — into an ice-cold tarn, black as a wolf's mouth. That everyone going to Gásadalur followed this same alpine route I did not find comforting in the least.

After I reached the top and began the descent, I did fall down — luckily I landed on the cushion of my rucksack. A brown and gray sheep materialized out of nowhere and regarded me with the same condescension as the Foula sheep, no doubt thinking that if I'd been allotted the normal four legs instead of two rather futile ones I'd be able to negotiate my way through the dips and depressions of life more easily. Even as it watched me, it continued to munch away on grass; there's so little sustenance at these heights that if a sheep stopped eating, it would die. The amount taken in must at least equal the energy expended to ingest it, which gives the sheep indefatigable jaw muscles and a tongue as strong as a bicep.

At last I reached the bottom and came to Gásadalur itself, a tight cluster of stone houses and outbuildings barricaded on three sides in the concavity of a deep valley and poised as if to plunge into Mykinesfjördur on the fourth. Shortly after I plunked myself down beside the single-pupil schoolhouse, I was surrounded by Gásadalur folk who acted as if they hadn't had a conversation with an outsider in ten years. Maybe they hadn't.

"You are an American?" asked an old woman. "*Já, já,* I was once four days in Liverpool and loved it . . ."

Said her husband: "Following the Vikings? *Já, já.* There was a Viking named Grim Kamban who lived in this very valley. He would slit a sheep's throat, put his lips to the blood, and pound the ribs with his knees to get the last drop. A fine polite man he was, and my cousin."

Their son crawled out from under a quadriplegic tractor and said: "I am reading in Danish translation your Zane Grey. Next I will start in on *The Hite Report.* So much literature comes from America."

"Grim Kamban had his house beside Heinanøy," said the old woman. "He had one good eye and one bad eye and his wife's name was Tove."

"*Nej,* woman," interjected her husband, pointing due east. "Grim lived *there* and his wife's name was Ingrid."

The son said: "Grim Kamban never got married. That is all I know about him. By the way, do you think I should read the new Harold Robbins?"

The husband: "Grim was a sorcerer who could bring you good weather or bring you bad weather. No one who lived after him could bring any sort of weather at all."

Said the son: "Do you know that a demitasse of sperm would impregnate all the women in the world? I have read it in a book."

A while later I walked to the cliffs where I sat down and ate some scurvy grass (*Cochlearia officinalis*), a salt-blown plant whose every leaf is said to contain as much vitamin C as an orange. I noticed a man with a fowling net crouched in a sort of blind below me. He had trussed up a bunch of dead fulmars for the inspection of live ones, who hovered around these decoys as if they were envious of this splendid new nesting site (which, of course, was Death itself). The man snagged two birds at once with a sudden jerk, getting one on the wing and batting the other to the ground, where it stumbled around like a drunken, overweight dowager. On the ground the fulmar — a very primitive bird, which can't swim or dive — can barely walk, which it compensates for dramatically in the air, bucking even the fiercest of gales. The man took this floundering bird and deftly wrung its neck, then trussed it up with the others, each of which would bring him fifteen *kroner* on the Torshavn market. The Faeroese have a grand passion for fulmars — why, I'm not sure, since the flesh of this extraordinarily oily bird tastes a little like an old tire doused with kerosene.

When he saw me, the man got up and offered me a thermos of coffee ten times stronger than truck-stop java. Then he began to tell me about his fourteen relatives in Minnesota, the computer whiz nephew in Copenhagen, and his mentally retarded stepson in the Torshavn lunatic asylum. He also told me about his second cousin Rasmus, who had once lived on Tindholmur, a nearby skerry with a spine like a stegosaurus. One day a sea eagle dropped down and carried Rasmus's two-year-old son to the highest pinnacle on the skerry. Rasmus and his wife got up there somehow, in the teeth of a winter storm, but when they reached the top they found their little boy dead from exposure. They were heartbroken, so they ended up in Torshavn, too. Nowadays Tindholmur was occupied only by a family of cranky and extremely antisocial trolls.

"Trolls? Do you believe in trolls?"

"*Nej*, but they're there anyway."

I walked on into the evening. The fog returned — a swirling, patchy, sunset gossamer fog. Four miles from Gásadalur I pitched my tent in the glacial litter at the base of a mountain called Eysturtindur, a setting so tranquil, so hushed, so silent that the only noise in all the world was the roar of vaporized fuel from my campstove, which sounded like a jet taking off.

June 13. Two days later I'm still in the same place, a charmless place, no doubt, but that's only one of its virtues, another being its handsome and single-minded uniformity. Each of the neighboring mountains has the same beveled Ice Age summit, the same cataracts, and has been eroded into precisely the same elephant's hide. The only wildlife is a pair of ravens who keep paying me visits on the off chance I have died and they can feast on my luscious eyes (ravens have astonishing sight, it's said, because of their fondness for eyes). The only flora worthy of the name is a lone cushion, perhaps a hundred years old, of moss campion. Other than that, there's not a shred of pastoral softness here, not even an errant fiber. No trees, hedges, shrubs, vines, crabgrass, or fragrant flowers. Yesterday I climbed the scoria-ridden slope of Eysturtindur and cast a line into Fjallavatn: no fish, either.

I'm installed in a land that's dead with the absolute death of the stillborn. Even movement seems to be atrophied: a waterfall hangs from Artindur like a narrow scarf of lace, its vapory dust a frozen painting. By all rights I should be feeling a little lonesome for my own kind and kindred, but I'm not. Instead I'm possessed by a sort of neolithic love, wherein familiar objects take on richer colors, or richer noncolors, and my thoughts new, more canny edges. It occurs to me that I'm seeing an irreducible landscape where not variety — the facile handmaid of the tropics — but the lack thereof presents itself with utmost simplicity. The planet might have looked like this when its crust was first squeezed into shape. I feel as if I'm peering at the first till, the first talus, the first waterfall, the first larval certainties, and the first lichen, ever. I scrape up some of this lichen — scraggly, lusterless map lichen — from a piece of porphyritic rock and regard it admiringly. It is hard not to admire something that requires only a few nourishing motes of dust, a drop or two of water, and a few stray beams of sunlight to be perfectly content for fifteen or twenty years at a whack.

I've read a bit of Epictetus and sponged out the condensation inside my tent. This morning I dug around in a nearby mound of rocks that looked man-made, on the whim they might be archaeological, something thrown together by the long-departed Grim Kamban. But all I found were the dead husk of a fulmar and an earthworm. *An earthworm!* Somehow I never would have imagined earthworms in Faeroe . . .

After I left Gásadalur, I took the mailboat *Solan* (Gannet) from Vágar to the island of Mykines, ten miles out to sea. Tidal currents make navigation hard in Faeroe, and nowhere harder than between Vágar and Mykines, where the currents range between seven and eight knots, with the wind usually blowing in the opposite direction. Sometimes it's impossible to make a landfall on the island on a lovely summer's day, even when there's only a light breeze from the south. Mykines folk used to have a ready-made and attractive explanation for this: the sea is unusually high-strung and it shakes with excitement every time a boat slides over it.

Whether this far-flung, westernmost island in the archipelago was ever a Viking sanctuary remains, like most early Faeroese history, something of a mystery. The first settlers were concerned with hiding out or laying claim to some tract of inviolable space, which did not make them eager to chronicle their whereabouts. A few old stories refer not to Vikings but to Irish monks. That figures; medieval Irish monks tended to construct their stone hives on obstinate rock, choosing the most remote and godforsaken islets, like the Skelligs, for their god-fearing devotions. Indeed, the founding myth of Mykines is not unlike the founding myths of certain Irish islands, particularly Inishbofin, where I once was holed up for an entire Force 9 winter. Inishbofin (so the story goes) was a magical place, hovering just above the sea, until a fisherman dumped a pipeful of live embers on its shores and ended the enchantment. Mykines, too, was once a hovering island, but then a rather myopic fisherman mistook it for a whale and flung a lump of steaming ox shit at it. The island lost its magic and also acquired its present name, which means "Ox Shit Promontory."

During my first visit to Faeroe I'd stayed on Mykines for two days and decided that life on the island could not possibly continue. It was the same old story: the elderly people were dying off and the young people were moving to Torshavn and not coming back except on holidays. The school didn't even have the one pupil Gásadalur can boast; I recall looking through the window and seeing an incredibly forlorn stuffed puffin on the teacher's desk. In 1970 automation of the lighthouse had done away with the last officially sanctioned jobs on the island, which left only the tending of sheep, parochial work nobody seemed to appreciate as much as boogying in the Kremlin Disco in Torshavn.

On my previous visit I'd met a rumpled sixtyish man named Oskar whose back was bent nearly double from hauling puffins, often a hundred a day, sometimes two hundred, from the fowling cliffs to the village. You had to have a bent back on Mykines, he told me, or you weren't considered much of a man. I asked him if he thought the island had any sort of future. He finished smoking a cigarette and crushed it under his shoe. "Does that cigarette have any future?" he said, a cough punctuating his words.

Upon my return, the first thing I did was visit Oskar's red corrugated-iron house on the edge of the village. From his daughter-in-law, who had the house for the summer with her husband, I learned that Oskar had been dead for nearly five years. Lung cancer, I figured. I was wrong. He'd been climbing the steps from the pier and slipped on some seaweed and toppled back to the bottom again, breaking half the bones in his body. It was ironic that a man who'd spent his life scaling breakneck bird cliffs should be done in by a couple of concrete steps. But then I'd once attended a lecture by Sir Edmund Hillary and seen the famed Himalayan climber stumble and fall down while attempting to mount the four steps up to the podium. Apparently those domestic steps were far more treacherous for him than the snow and ice cornice on the upper summit ridge of Everest.

After a day on Mykines, I changed my mind about life not going on. A sort of life *was* going on, beating with a reasonable version of a pulse, but that life consisted for the most part of travelers like myself. There were maybe a dozen of us — one third of the island's population. Our tribe could only increase as the Mykines tribe dwindled away, a few falling down steps, most simply emigrating, until there would be, sad to say, only our peripatetic selves. We were the future of all places condemned by remoteness to a lingering, photogenic death.

Almost all of us had been to the ends of the globe and had brought back tales to tell around the campfire. There was the German who had been bitten on the thigh by an elephant seal off Tierra del Fuego; he dropped his trousers for us and proudly displayed the big rosy scar. There was the Swede who claimed he'd learned to hear the aurora borealis ("a faint rustling sound") by camping among the sharp-eared Lapps in the far north of his homeland. There was the Dane who spoke knowledgeably about the African method of calculating distances by the number of cigarettes smoked on a journey. And then there was the other Dane who'd brought back a Gurkha Kukri knife from his travels in war-torn Afghanistan. It could cut a man's throat so clean, he said, that the victim would think the knife had missed — until he smiled. I borrowed the knife and used it to cut my *skerpikjøt* (wind-dried Faeroese lamb), which is twice as tough as Canadian pemmican or South African biltong. I can't vouch for throats, but on the *skerpikjøt* the knife did rather well.

The second day I walked around the island with a callipygian Swedish girl named Gretta. She had a warm-eyed and frisky manner which carried the hint that anywhere else in the world she'd throw off her clothes and romp nakedly in the woods. But Mykines had no woods (no trees, either) and all she wanted to do here was gaze ecstatically at seabirds. And gaze at them she did. We both did. We watched a group of puffins radiate like streams of tracer bullets from their respective burrows. After landing, they'd flick their bodies around jerkily, like Kabuki actors. I saw one puffin execute a nearly perfect 360-degree swivel of its head, reminding me of the puffin I'd killed on Mingulay, which had turned around casually to watch me strangle it.

At one point Gretta looked sweetly into my eyes and said, "Have you puffins in your country?"

"Only a handful, in Maine."

"Too bad. I will never go to America."

We edged down a narrow path sliced out of the cliff face and then crossed a chasm spanned by a lolloping bridge to Mykinesholmur, a small island separated by a twenty-foot gut from Mykines. As we passed a grassy outfield, a pair of oystercatchers suddenly burst into hysterical life, pip-pipping at us like two Sunday school teachers outraged by a classroom blasphemer. Indeed, the oystercatcher is such a pip-pipping nuisance the Faeroese have made it their national bird in a futile effort to get it to shut up.

Finally we stood at the huge gannet colony that was Mykinesholmur's pièce de résistance. Hundreds of these big swan-like birds wheeled around above us like snowflakes in a snowstorm, moving steadily northward overhead, then southward to complete the circle. Every twenty seconds or so one would leave the storm and sail close to have a look at us with its cold orange eyes. Apparently finding us uninteresting, he would sail off to join the rest of the snowstorm. Below us was the gannet equivalent of the Mong Kok section of Hong Kong — thousands of nesting pairs hemmed together in a stifling urban ghetto, alternately conversing and arguing with each other in their hay-cutting-machine guttural. Each pair had its own high hummock of grass and seaweed to keep the chicks well above the rising tide of guano, a good strong whiff of which, mixed with stale fish and dead birds, drifted up and nearly knocked us off our feet.

Anyone who believes in the infinite grace of birds has probably never watched a gannet trying to land on its nest. It comes down out of the sky as gracefully as a descending boulder, braking with wings and broad feet splayed out and, I'm convinced, eyes tightly shut. Usually it overshoots its nest like a plane that's overshot the runway, ending up on a neighbor's hummock, whereupon it's subjected to a barrage of angry beaks and irate cries until it manages to lurch back to its own — wait! — not its own but another angry neighbor's hummock, at which point the same thing happens again.

While Gretta took some pictures of the gannets, I wandered off and met an islander scything hay, who looked like he'd rather talk than work. He was a man at once raw-boned and burly, a sort of cross between Hans Christian Andersen and Jackie Gleason. He spat out a wad of phlegm and told me this story:

Once upon a time there was a giant named Torur who owned Gásadalur. Torur wanted to own Mykines as well, so he hopped over in a single bound. An islander named Olí saw him coming and drew out his knife to cut Mykinesholmur off from Mykines. But anyone who could hop over from Vágar could hop this chasm easily, which Torur did, saying: "Give me Mykines!" Olí gave him only the taste of his fist. And the two of them did battle for days, perhap even months, until Olí picked up his knife and gouged out one of the giant's eyes. Now Torur pleaded for mercy and promised Mykines three priceless gifts if his life were spared; but since every such promise has a catch, he added that nobody could laugh at these gifts or they'd be withdrawn. Thus it happened that a big whale washed ashore next spring, but as this whale had only one eye, like the now-humbled Torur, everybody had a good laugh at it. No whales have washed up on Mykines since then. Next an enormous piece of driftwood washed ashore, but as this driftwood seemed to have the face of a prominent Faeroese politician, everybody laughed at it, too. Now it's impossible to find driftwood on Mykines. But then a migration of gannets came to the island and it put people in such awe that they forgot to laugh. Soon all the girls in Mykines dressed in white frocks and danced along the cliffs. And every spring after that they donned white frocks and danced to celebrate the gannets' return from the sea.

"Did you ever see them doing that?" I asked him.

"Many times. It was the sweetest sight in all the world. But now there are no more girls on the island." He swallowed these words like a quick shot of whisky and then resumed his work.

It happened that this voluble haymaker was the lay preacher on Mykines. Later, when I saw him in the village, he invited me to his church. Now my religiosity is at best a bit suspect, as I tend to believe in all the old bugbears and superstitions the early missionaries thought they'd wiped out. But I figured that anybody who could speak with such authority about one-eyed giants and dancing girls ought to be able to preach a pretty good sermon. Sunday morning I duly showed up at the closet-sized church. I was the only one; the preacher and I waited for half an hour, but the rest of the island seemed to be sleeping in or fishing out, so I was forced to assume the role of congregation. The preacher read to me from the Scriptures, though he kept losing his place, stumbling over words, and dropping his glasses at key moments. Then we slogged through a couple of hymns, he in Faeroese, I in Icelandic (I can make myself understood in Faeroese by speaking this kindred tongue), while both of us looked yearningly out the window. Toward the middle of the service I realized that the Lutheran faith permitted only genuine men of the cloth to deliver a sermon. Thus I was to be deprived of a moral lesson on the importance of dancing girls in God's scheme of things. I was somewhat disappointed, but I did like the man's pulpit style, which reminded me of a foot stuck resolutely in the wrong shoe. He closed out the ceremony with a hymn, singing the words in a suddenly animated falsetto that I could not help but attribute to a hangover. Afterward I filed silently down the aisle and out the door, pausing only to inspect a pair of candlesticks made of basalt, a far more appropriate material for Mykines than brass.

That evening I sailed back to Vágar content to know that at least the spiritual life of the island was in rather good hands.

A day or two later I was walking north of the Vágar airport when I noticed quite a lot of commotion on the road. BMWs and Mercedes rocketed past me with big nasty grappling hooks called *soknaronguls* sticking out their windows. Such weaponry could only mean a *grindadrap*. I hitched a lift with a man and his *soknarongul* to Midvagur, where a big crowd had already gathered on the shore, including the entire population of the local old folks' home, some in wheelchairs, others propped up by their nurses, still others scarcely sentient.

This was a supreme bit of luck. Midvagur is possibly the best whaling bay in Faeroe; if not the best, it's certainly the gentlest, for its sloping shoals allow whales to beach themselves. Off a rocky eminence like Tindholmur, they'd have to be killed from boats after a drawn-out chase, a procedure that Faeroese don't like any more than the whales do. For a chase builds up fear in the pod to a near-fever pitch and, like most hunters, the Faeroese think the rush of adrenalin caused by fear toughens the meat of their prey.

Looking out at the slate-gray sea, I couldn't see anything except billow after billow of fog cascading in like thickly meshed clumps of wire. I did hear the pulse of motors and echo sounders switched on to drive the whales shoreward. Suddenly the fog lifted for a moment, and there were the boats and then the whales raising little cockatoo crests of foam in the sea. The boats had fashioned a tight knot around them and were shepherding them toward us. The *grindaformenn* was directing the action through a megaphone, calling out something that sounded suspiciously like — he was still a good distance away — "Have a good day! Have a good day!" Later I figured out he was actually asking for a sort of harpoon called a *hvalvákn*.

It was a genuinely eerie scene — the thrashing black tails and ghostly boats emerging and then being swallowed up by the fog, easing closer and closer to shore. Finally the foreman plunged his *hvalvákn* into the bellwether, and a crimson fountain of blood shot up at least two feet into the air. The wounded whale's high-pitched, almost piglike squeal was picked up by some of the other whales, who echoed their leader's pain and confusion in comradely fashion. This initial blow was significant, since most whales, and especially long-finned pilot whales, seem unwilling to swim free once the first blood has been shed. They swim away briefly, but always return with suicidal persistence to the bloody scene, their herd instinct so powerful that one whale's anguish is every whale's anguish.

All at once a collective shout went up from the men on shore, and like a banzai charge they leaped into the frigid water, armed with knives and *soknaronguls.* Most of these men seemed to have been snatched from work; many still wore their street clothes, including one man, snatched from perhaps more elevated work than the others, who was wearing a suit and matching tie.

The men splashed over to where the whales were beaching themselves. Each gaffed a whale and then thrust his ten-inch *grindaknivur* through that whale's spinal marrow, just beneath the crescent blow hole. The whale would thrash violently, breaking its own spinal cord, and die. Two such thrusts by a master can do in a whale almost as fast as you can read this sentence; one thrust of a whale's fluke can do in a man, rendering him a species of jelly. But the only casualty I saw at this *grindadrap* was one of the boats, whose planks exploded on contact with a bull's tail. The men from that boat just swam through the teeming whales to the shore, grabbed knives, and splashed back into the water to join in the killing.

The whales' blood was warm and the water temperature quite cold, so the blood stayed on the surface and filled the entire bay, which turned into a pool of reddish froth. The air had the aroma of a surgical ward. At one point I saw my friend Pinnock in the process of making a video of this spectacle. Such a video would assume near-porno status in his native England, where nonhuman creatures tend to be more highly regarded than human ones and the idea of hunting derives from an aristocracy that never had to eat (haunch of fox?) what it killed.

"Disgusting," said Pinnock.

"Yes," I said. "But for God's sake, don't try to change it."

"I actually think these blokes are enjoying themselves."

"Of course they are. It's not every day you can get a year's supply of meat with a few quick knife thrusts."

The children in attendance indeed seemed to be enjoying themselves. A couple of them were playing with the hearts and kidneys of whales already cut open to keep the meat from spoiling. Little eight- and nine-year-olds trying to pick up a great sloshing whale heart! Little six- and seven-year-olds attempting the kidney toss! At an age when most American kids still swoon at the thought of Mickey Mouse, these children were gaining a familiarity with the shape and feel of innards. They were learning that their food was not created mysteriously in the abysm of some supermarket, but came from a grandiose living organism that had to be hunted down and put to death if perchance they wished to eat. Heaven help them if they refused this learning. Just about all they'd have left would be glitzy automobiles and a couple of antisocial trolls . . .

By evening the carcasses of ninety-eight whales had been gaffed to shore and laid side by side like casualities of war. They would spend the night there (who'd steal a multi-ton whale?), and come tomorrow they would be flensed and the meat apportioned to all the people in Midvagur. In principle even a newborn baby can apply for its own small piece of meat.

That night I went to the *grindadansur* held in the Midvagur recreation hall. It had begun just before midnight; I arrived a little late, so I missed the argument about whether Led Zeppelin was appropriate fare for a whale dance. Apparently it wasn't, because by the time I got to the hall, I heard nothing so artificial as prerecorded vinyl or so impertinent as instrumental accompaniment. This was dancing in its antique simplicity, borrowed from medieval Europe and never returned, even as medieval Europe had evolved into modern Europe and as the fox trot had become a bit archaic. It was not very different from the silver wedding dance I had taken part in a couple of years back except that some of this crowd still had random streaks of blood and gore on their clothes, a feature I did not recall from the earlier shindig.

The entire village joined hands in a circle and moved two steps forward and, with a little kick, one step back, stomping the wooden floor with salvos of cannon shot. As the circle twisted and turned into narrow bights, like an injured snake, the villagers belted out the "Grindevise" (". . . to kill a whale . . . that's our fondest desire . . .") in a kind of rough chant that suggested a heathen prayer to the gods of fog and gray weather. They followed this up with a two-hundred-verse ballad about Sigurd the Dragon Slayer, recounting Sigurd's exploits forest-dark centuries ago as if they'd happened just that morning in a Midvagur moraine.

I had sprained my ankle on Mykines, and after an attempt at Sigurd's heroic deeds that sent jolts of affliction as far up as my thighs, I elected to save my ankle for assaults on Icelandic lava. I sat down in a corner with a group of men who were getting very, very drunk. We talked; more precisely, I talked and they drank. At a break in the dance a hefty young woman who seemed built to withstand northern winters sat down breathlessly next to me, saying that there had already been four heart attacks this year alone at Faeroese dances. *We* talked. She wondered what I thought of these curious, rock-ribbed, foot-stomping islands. I said I liked them just fine, but the place I especially liked was Gásadalur, to which I could imagine myself repairing, worldly goods in my rucksack, if the heat were on in my own cluttered part of the world.

"But haven't you heard?" she said. "Last week the Løgting allotted forty million *kroner* for a tunnel to Gásadalur. It will go through the mountain, Knukarnir, that you climbed."

At this gloomy piece of news I realized it was time to move on to the next last place.

Chapter 4

NAKED CITY

IN TORSHAVN I booked passage to Reykjavik, Iceland, on a Faeroese cargo vessel. The captain led me to a cabin next door to the vociferous heartbeat of the engine and said it was the best accommodation in the house. Not only that, it was the only accommodation unless I wanted to spend the two-day trip alfresco, embracing the wind, sleet, and rain of the open deck. I got the distinct impression that he would have preferred the deck, along with its surly elements, himself. He looked like Coleridge's Ancient Mariner, except for the tell-tale albatross; he'd skippered this boat more than a hundred times between Faeroe and Iceland, one of the most raucous weather belts in the world, and the route had etched itself permanently in the lineaments of his face.

Halfway to Iceland we were struck by a nor'easter barreling down from the Arctic. Squalls began snatching sheets of hissing water off the sea and hurling them to leeward. Lumpy waves raced after each other like frightened horses. I'd been playing chess with one of the crew on night watch, and I first noticed the effect of these waves on the chess pieces, which kept making moves of their own accord. Soon I heard the slithering and lolloping of loose objects as they gravitated temporarily to new places of rest on the deck. Phantoms opened and slammed doors and made eerie clanking sounds in the hold. First I grew indifferent to the game, then I couldn't locate its whereabouts; for while my mind and eyes told me I was in one place, my inner ear told me I was somewhere else, far, far away from the tidy squares of the chessboard. Dizziness grabbed hold of me. I had to excuse myself and take a turn at the gunnels, bringing up (as old sailors used to say) Jonah, who in this instance was a meal of boiled haddock, stewed turnips, and rhubarb pie. Afterward I proceeded to my cabin, where I lay inert and stupefied while steep seas the color of cappuccino played with the boat like a cat with a mouse. That's a tired phrase, I know, but from the limited perspective of the mouse it seems entirely original.

Of all the miseries the sea can inflict on the hapless sailor — submerged reefs, shipwreck, giant waterspouts, hurricanes, impressment into a foreign navy, amputation without anesthetic, pirates, equatorial malaise, hijacking, icebergs, attacks by torpedos or multitentacled monsters of the deep — the one responsible for the most suffering is *mal de mer*. At best you feel like you're going to die; at worst you feel like you're *not* going to die but live on forever, besieged by dry heaves and cascades of sweat, never to reach the calm of port again. Inevitably, scenes from your past life float before your eyes just as if you were drowning, but they're mixed up and fragmentary, like a badly choreographed nightmare rather than a settling of old accounts before the final fadeout. Sometimes your friends and loved ones visit you in your delirium, but they're seasick, too.

I'd always thought I was immune to this malady. Always thought I could weather any storm, no matter how disoriented my inner ear. Always identified with those old ladies who say that on their last sea voyage they were the only ones capable of sitting down to dinner with the captain. But I wasn't immune. Neither, in fact, was our captain, in whose face shades of green and yellow fought for supremacy. At one point he came down and informed me that he'd been seasick in one form or another virtually every day of his nautical career. I asked why he didn't get a landlubbing job and spare himself the agony. He loved the sea, he said.

I tried to sleep on my back to favor the grim conditions inside my stomach, but each time I lay on my back I was flung to my stomach by the swivel-and-jive action of the boat. By now the crew were coming down to offer me selections from their repertoire of seasickness cures. In the old days you'd have your choice of morphine with atropine, a belladonna plaster on your stomach, a Worcestershire sauce cocktail, irrigation of the ear canals, or a spinal ice bag. One bygone cure indicated copulation performed repeatedly, but that was clearly out of the question here, as was French author Blaise Cendrars's suggestion that the victim read Descartes (I had no Descartes).

The Faeroese were not necessarily more practical. One of them mentioned a holistic remedy from the Orient; I was supposed to rub the skin between my thumb and forefinger because this particular skin connected directly with the semicircular membranes that were causing all the trouble. The ship's cook told me I ought to chew a few limes, for limes make the throat contract and then you lose all interest (because you've suffocated to death?) in throwing up. But a quick check of his galley produced no limes. Another crew member gave me the very latest in Danish seasickness medication, a capsule which, unlike Dramamine, worked *after* you got sick. Up it came twice as fast as it went down. In the end I simply went up to the deck again to breathe in the fresh, violent air of the North Atlantic and gaze at the far horizon rather than the claustrophobic walls of my cabin. After a while I was able to discern the pyramid-shaped mountains of southern Iceland, which did make me feel somewhat better if only because it meant the journey was almost over.

We headed into the relatively calm waters of Faxaflói and at last dropped anchor in Reykjavik harbor. The boat crouched at the pier as if it were trying to get back its breath after the long hard struggle with the waves. The gangway was lowered, and I alternately limped and staggered, stumbled and crawled, crept and fell, ashore.

Only a chunk of lava's throw from where I came ashore, a man called Ingólfur Árnarson set up the first documented Viking housekeeping on the island. Others might have come before him, but the *Burke's Peerage* of Iceland, *Landnámabók* (*The Book of Settlements*), does not list them. It does list Ingólfur, who was forced to flee his native Norway after a murder conviction. At Anarhóll the town honors its homicidal founding father with a stern-visaged statue: the bronze Ingólfur peers forth from his pedestal as if in search of new worlds to settle, new men to bump off.

In the year 874 Ingólfur left Dalsfjord in western Norway, and upon seeing this big new island he threw his chair pillars into the sea. Chair pillars had elaborate mythological carvings and were by far the most important item of furniture in the Viking home. Wherever they washed up, that was where the gods decreed a man had to settle. Doubtless a lot of pretty awful places in Iceland got settled in this fashion. Ingólfur's chair pillars could have washed up in Skeidarásandur, a desolation of shifting sands in the southeast that only a skua — 3,000 mating pairs, in fact — could appreciate. Or they could have washed up at the infertile plinth of a bird-fretted cliff in the Westfjords. But Ingólfur was lucky, although his luck was a little slow in coming. It took him nearly three years to locate his chair pillars because he had no idea where the westward currents would dump them and also because he had to take time off to avenge the death of his blood-brother Hjörleif, killed by some uppity Irish slaves (early settlers often stopped off in Ireland to pick up cheap farm labor). Finally he found the pillars half buried in the sand on the shores of Faxaflói, only a short distance from the best anchorage on the entire southern coast. Here he settled, naming the place Reykur-Vik, Smoky Bay, in honor of the geothermal springs that sent big columns of steam into the air. Ingólfur must have been astonished; they didn't have smoky bays back in Norway.

Others followed Ingólfur's example, and by the year 950 the coastal fringes of the island were largely accounted for by his pillar-tossing countrymen, a testy, even solipsistic race of people who seemed to believe that good mountains make good neighbors and wide fjords make even better neighbors. The republic they fashioned was a fine mix of parity and ruthlessness. On the one hand, it was the site of the world's first democratic parliament, the Althing, established in 930; on the other, those first democrats used to salt the heads of their enemies and carry them around to show off to each other. Capital punishment was prohibited by law, but family vendettas of almost genocidal proportions (like the one in *Njal's Saga*) were perfectly legal. Literature prospered. *Most* literature, that is; perpetrators of love poems were given very stiff fines. Under Christianity, which arrived around the year 1000, heathen survivals flourished as never before; one family achieved particular notoriety by worshiping its stallion's penis rather than the Christian god (the church tolerated this as long as the family did not try to win converts for the stallion).

Something had to be tolerant, as the island's hyperactive geology was not. Since Ingólfur first cast his chair pillars, Iceland has erupted 125 different times from thirty different volcanoes. One in particular, Hekla (from which the word "heck" is possibily derived), has erupted sixteen times, burying up to half the island with ash and debris. Hekla's eruption in 1300 covered 30,000 square miles with ash, cinders, and tephra; in 1766 another of its displays lasted two whole years. But by far the worst volcanic outburst — indeed, the biggest eruption recorded anywhere, *ever* — occurred in 1783 when the twenty-mile-long row of craters known as Lakagigar blew forth ash cones, scoria, and lava fourteen times the bulk of Mont Blanc. Rivers were stopped up, grass poisoned, forest lands destroyed, and over one fifth of Iceland's population and four fifths of its livestock killed. At this point the Danish government (Iceland was then, like Greenland, under the proprietorship of the Danish crown) proposed that the much-erupted-upon survivors be removed to the peace and quiet of Jutland. Icelanders demurred. They'd come to love their autodestructing island as if it were the sweetest place on earth.

Curtailed by eruptions, hobbled by insidious weather, wiped out by family feuds, and kidnaped by Barbary pirates, the population of Iceland hardly changed for a millennium. Sixty thousand people lived on the island in 1850, the same number as in 950. In 1806 Reykjavik had only three hundred inhabitants, of whom twenty-seven were in jail for public drunkenness. The town had a single dilapidated church, shared by parishioners and ravens alike, with the latter constantly disturbing the devotions of the former with their expostulating croaks and droppings. The court of justice doubled as a tailor's shop, and the best house, the Danish governor's manse, had but six rooms. In 1809 the English botanist William Hooker reported that not a single vegetable was to be found in all of Reykjavik; most irregular, he noted, for a capital city. Half a century later the novelist Anthony Trollope visited Iceland and made a similar observation: there wasn't a single cabbage in Reykjavik (he didn't find one in Torshavn, either).

Nowadays Trollope would have no difficulty finding cabbages in Reykjavik or even something a friend of mine once referred to as a zucchini sasquatch, though by the time it reaches the shops, having been shipped from Europe or America, the sasquatch is often black inside. That's not all; present-day Reykjavik has exactly the same state-of-the-art, design-dripping gizmos you can find anywhere else, as well as the same rarefied bourgeois artifacts, from underwear handwoven by Tibetan virgins to computers that let you play rat-maze games on them.

Ingólfur Árnarson's Smoky Bay is now a modern metropolis of 110,000 people, half the population of the island. As more and more inhabitants of the outback decide to take advantage of the gratis steam heat, Reykjavik just grows . . . and grows . . . and grows, like a child with a ghastly pituitary condition. Each time I come here I discover whole new ferroconcrete suburbs, which look as if they were designed by the architect who designed Hitler's bunkers. Each time I think: *Enough!* If this madcap growth keeps up, the future of Iceland will consist only of Reykjavik and lots of memories. Yet. And yet. Beneath all that concrete you can still hear the pulse of Ingólfur's original thermal springs.

July 4. Stuck in the Salvation Army Hostel for a while. My ankle's still bothering me, and I may have come down with a mild case of pneumonia from exposing myself to the hostile elements on that cargo ship. It's no consolation, but the other folks here don't look too good, either. Most of them are sea dogs drying out, and a couple are madder than March hares. There's a violently anti-American dwarf named Bjóssi and a black gentleman from Ethiopia who seems to consider himself Icelandic by adoption if not by birthright, having been baptized back home by an Icelandic missionary. I'm drinking a nonalcoholic lager called Egil Skallagrimsson because the Salvation Army doesn't tolerate consumption of the real thing on its premises. Egil Skallagrimsson was a Saga hero renowned for the splendor of his booze-ups. To name a watery beer after him is a bit like naming a chocolate milk after King Lear.

One of my fellow inmates is a fisherman named Indridi

(Lonesome Rider), who, like nearly everyone in Reykjavik, comes from somewhere else, in his case the now empty, mountain-spined peninsula of Hornstrandir in the Westfjords. Indridi stands about six foot four, and his head appears to have been cut from a Mesozoic boulder, even though it sports a Daytona 500 haircut. Hornstrandir people seem to have a special toughness about them, as if they got their milk from sucking the stones of talus slopes. Indridi says his father was once buried in an avalanche and amused himself until help came by composing forty-seven *slettabönds* (a *slettabönd* is a rigorous four-line verse form that has the same meaning backward and forward). His mother, he says, gave birth to sixteen offspring out of a deep-rooted fear that the Icelandic race would die out. Indridi himself takes snuff from thumb to elbow, throwing back his head and inhaling the dirty brown powder with a single elephantine snort. It's a thing of beauty to watch him do this, yet he warns me that snuff-taking is not without its perils. His father's eldest brother died as a consequence of it. Cancer? I asked. *Nej, nej,* he replied. The old boy was standing in his open lugger off Ólafsvik in Snaefellsnes and threw back his head so fervently he pitched himself forthwith into the sea and drowned.

The dwarf Bjóssi isn't the only xenophobe billeted with the Salvation Army. There's a full-sized xenophobe who admits to eighteen names, one for each of the crew who went down with a Westfjord fishing boat the day before he was born, but "I am called Baldur, for short." Baldur does not love his native lack of soil so much as he detests and abhors its foreign mercenaries, particularly the American variety; the American NATO base at Keflavík he likens to Nicaragua and Grenada (in loftier moments he likens it to Vietnam). A few years back he helped Sveinbjörn Beinteinsson, the high priest of the Icelandic heathen revival, raise a *nid* pole with a horse's head on it against Keflavík's F-3 Phantom flyers and nuclear-capable P3-C surveillance planes. The horse's head did no good whatsoever, Baldur says. Perhaps Icelanders need to use something a little more high tech, like a cobalt bomb . . .

One night Baldur stumbles in, drunk of course, blinks his eyes, and, not recognizing me, demands to know where I'm from. Since I don't really want to get into another Keflavík dispute (I agree with the anti-Keflavík faction anyway), I reply, "From Mo-o-o-orman-n-nsk, in the USSR." The Soviet four-diesel fishing vessel *Surikov* is presently docked in Reykjavik, so Murmansk doesn't seem that far-fetched. What I didn't expect is that Baldur has been on an Icelandic fishing vessel that actually docked in Murmansk. He knows the town rather well and wants to find out which part my digs are in. Before I can answer, he reaches out and grabs my arm with a vise-like grip. Damn, I think, now he's going to punish me for not being from Murmansk. Instead he says: "You will jump your boat. You will get political asylum in Iceland. You will stay with me, your friend, Baldur." Relieved, I continue the ruse: "*Njet! Njet!* That I cannot do, Baldur. I have a wife and four little ones in Murmansk. The KGB would impound them if I jumped my boat." "Do not worry about your family, poor man. I, Baldur Magnússon, will write a letter to the KGB and they will ship your wife and little ones to Iceland. If they do not, then they are in big trouble." He waves a menacing fist in my face even as he continues to clutch my arm. "Maybe this political asylum," I suggest, "maybe it is for ballet dancers only, not fishermen . . ." "*Hver fjandinn! Fishermen must have freedom, too!*" "Sh-h-h!" I whisper. "I think that man over there" — I point to Indridi, seated happily in front of the TV weather map — "is KGB." "*Freedom is for everyone!*" Baldur shouts again, reiterating his island's raison d'être in Viking times. And I'm certain he would have force-fed me this freedom had he not toppled soddenly to the floor on his way to punching out Indridi.

For four days I was holed up in the Salvation Army Hostel, recuperating from my ills. By Saturday night I felt good enough to join the rest of Reykjavik in its weekend bacchanalia.

As the government tends to run a gulchlike deficit, all week Icelanders attempt to leapfrog the peaks and stately massifs of their island's inflation. Some work double and triple overtime. Others work at two jobs, like a heart surgeon I've heard about who also extracts worms from cod in a fish factory. Still others ride the bucking-bronco decks of North Atlantic trawlers, only to return home and learn that the bottom has dropped out of the Nigerian or Portuguese fish market. Saturday night turns all of this into a distant memory. People hit the streets (indoor drinking, in Iceland, is considered dissolute) like the prisoners in Beethoven's *Fidelio,* who, upon emerging from their dungeons, sing, "*O, welche Lust!*" — O, what joy! In Reykjavik, they're singing, "O, what lust," too.

By ten-thirty P.M., when I hit the street, the town center was already surrounded by a tight girdle of cars. They moved along almost imperceptibly and honked their horns like manic geese. Nobody was looking for a parking place or hoping to move forward any more than they hoped to levitate upward in this clutch of automotive inertia. They were just celebrating togetherness, Icelandic style. Celebrating a bumper-to-bumper intimacy wherein half the nation might be stuck in one monolithic traffic jam (the country that's congested together stays together).

I walked into Austurvöllur, the square built on the site of Ingólfur Árnarson's original home field. It was currently being occupied by the Salvation Army sea dogs, who were passing around a bottle of *brennivín,* the local anise-flavored schnapps, along with a bottle of vodka which I'm sorry to say was mixed with Coca-Cola:

"*Já, já.* Tycho Brahe died. His bladder exploded while he was waiting to meet the king of Denmark."

"Things like that will happen. If you have a bladder."

"Kindly do not explode your bladder around me, Jóhann."

"I'm Haukur. He's Jóhann, son of Bjartur, grandson of Egill, great-grandson of Thorgeir. The only alcoholic Stalinist in the country."

"My mother was a Stalinist. If I didn't support Stalin, I'd be insulting her memory."

"How's your bladder?"

"Full. *Skál!*"

"Skál!"

"Better the little fire that warms" — raising the *brennivín* bottle — "than the Great Fire that burns. *Skál!*"

Near the telephone exchange a dozen teenage boys were impaling a teenage girl on top of a stop sign. She was doing her best to help them. While they pushed, she was shinnying up the pole. One of the boys, whose haircut proclaimed him the northernmost of all the Mohicans, offered me a swig of *brennivín.* "Cowboys like candlelight and lacy things on ladies," he observed, quoting from some American pop song. It was apparently the only English he knew. Meanwhile the girl had made it to the top of the sign and was sitting astride it, savoring the moment like Hillary on the summit of Everest. I couldn't help but notice that she wasn't wearing any panties under her leather mini-dress.

"Tonight," the girl said, pointing down at her friends, "I will sleep with *you* and *you* and *you.*"

Now the motorcycle gang Sniglar (the Snails) fell in with the bumper-to-bumper traffic; they'd just returned from an international conference of motorcycle gangs in Copenhagen and seemed eager to show off new methods of making their bikes backfire. More horns honked. More drunks drank. I kept running into old acquaintances who didn't recognize me and total strangers who did. A bearded man who looked like Walt Whitman on a binge asked me what I was doing on board his trawler (he *did* seem to be bobbing up and down on the high seas). When I said this was downtown Reykjavik, not, unfortunately, his trawler, he gazed around in amazement. "*Allt er á tjá og tundri,*" he told me. Everything's topsy-turvy.

Everything *was* topsy-turvy. I attributed this at least as much to the Midnight Sun as to inflation or alcohol. From this Sun's spell no one is truly safe. It doesn't nod off like a normal sun in more conventional latitudes, though it doesn't linger in the sky like the fiery nocturnal orb of legend, either. It just settles comfortably on the horizon and casts off an invigorating dove-gray or purplish light. A person touched by this light feels in the grip of magic, as if he can do anything, even make the first ascent, hitherto thought impossible, of the difficult north face of a stop sign.

I joined a queue in front of Odal, a drinking establishment with a less rigid dress code than some of the others in town. I was able to get in after a twenty-minute wait and went immediately to the bar. A fellow of twenty or thereabouts was trying to impress a girl with a joke, only the end of which I caught: ". . . 'the best snuff I ever tasted,' the Hafnarfjördur man said. 'Well, it's not snuff,' came the reply, 'it's dried horse shit I put through my wife's coffee grinder.' " This was a Hafnarfjördur joke (the Icelandic equivalent of a Polack joke, named for a small town south of Reykjavik) and the girl thought it so funny she spilled half her drink on my hiking trousers.

The man to my left offered me his handkerchief. "*Thakk' fyrir,*" I said, daubing at the stain. "*Minnstu ekk a pad,*" he replied, and bought me a double *brennivín,* a rather kind gesture, I thought. The man wore a hand-tailored three-piece suit and silk ascot, and from his general demeanor I might have taken him for the Belgian ambassador. But after a bit of preliminary chitchat, he said:

"I am a murderer . . ."

Confessions like this have a habit of stopping conversation dead in its tracks. What can you say in response? Finally I replied, "So was Ingólfur Árnarson. Eric the Red, too."

"It's true. I killed my wife."

The calm, matter-of-fact way he spoke made killing one's wife seem like quite a reasonable proposition. He said he'd flung a hammer at her and it had struck her in the occiput. They were getting ready to leave for a dance in Kópavogur. She'd died instantly. Then he went on to tell me that he was presently "on leave" from the prison. The Icelandic prison system was very tolerant, he said; no matter what you did, whom you killed or mutilated, you could always leave the prison confines so long as you reported your whereabouts to the police every four hours.

"How many murderers are on the loose now?" I asked him.

"Oh, five or six." In his voice I could almost detect a hint of melancholy. After all, there was the robust example of the Sagas breathing down his back. In my Penguin Classic paperback of *Njal's Saga,* ninety-four killings and/or murders befall 355 pages of text (this body count doesn't include the casual head loppings and so forth neatly summarized with phrases like "Many men died . . .").

I was curious about prison life. The man told me that most inmates had their own keys and could come and go as they pleased. On weekends the prison was nearly empty. An inmate's only responsibility beyond the check-in calls was not to lose the key. *Not to lose the key,* the man repeated solemnly, and then remarked on the bad luck of a fellow inmate who happened to lose his key and was forced to spend the night in a hotel.

"In my country, we don't think it's right to kill people just because they've killed people. But in your country I would have gotten the *raffmagnsstoll.*"

There is no *raffmagnsstoll* (electric chair — literally, "amber power stool") in Iceland. Crimes seem to be looked upon as expressions of a strong national spirit or simply familial mischief making writ large. Father Justice indulges his wayward sons and daughters and, however outrageous their crimes, always seems to welcome them home again. A case in point is a man named Helgi Hóseasson. A fervent atheist, Helgi is famous for his efforts at calling attention to an injustice, namely, his baptism (a slap in the face to his deeply held sense of nonbelief). On a live TV broadcast a few years ago, he crept up behind the bishop of Iceland and the president and doused them both with a large vat of *skyr* (yogurt). He was sent to jail for a day or two in deference to his crime. Upon his release he painted the government office building with tar, for which he was reprimanded and told not to do it again. He didn't do it again. Instead he escalated his arsenal from yogurt and tar to kerosene, and burned down the Breiddalsvik Church, the second oldest church on the island. He got off after an inquiry seemed loath to prove his guilt.

My companion did not appreciate Helgi's rebelliousness. He said, "That man Helgi Hóseasson should be put away. He is a menace to our society. One day he will burn down the Althing."

"Possibly," I said. "By the way, how did you happen to murder your wife?"

He looked sheepish. "Too much *brennivín,*" he said. Then he asked if I'd like another double *brennivín.* I declined.

Music was going full blast in the background, and I listened to a song by the rock musician Magnús Th. Jónsson about a huge cod windfall. Magnús is a rocker partial to the High Baroque who likes to weave references to Hallgrimur Pétursson (1614–1674), poet and hymn writer, into his song lyrics. His music is quite popular with the young, though not perhaps as popular as a band called Sjálsfroún (The Masturbators), who aren't interested in the Baroque at all, but are interested in sex and violence. I once met Sjálsfroún's lead singer at a party and he said his life's dream was to visit Times Square some day.

As I moved around the edge of the dance floor, I picked up stray bits of conversation:

"I saw my first tree and was terrified . . ."

"Worse than your first ghost?"

"*Já, já.* Much worse."

"It was a family where everyone composed rhymes. Everyone except Inga."

"I know. She married an American."

". . . stuffed his favorite dog and gave it the place of honor in the living room . . ."

". . . tacked the pelt of his favorite horse to the wall . . ."

"*Já, já.* My first ghost scared me. Then I grew accustomed to them. One must."

Now I saw an old friend of mine, a filmmaker named Páll Steingrimsson, who had appeared on the TV show *To Tell the Truth* back in the 1960s (the contestants had to guess which of three candidates was an Icelander). But before I could walk over to him, an immovable bulk blocked my way and identified itself as Einar "ship builder, fish salter, taxi driver, seal hunter, and ski instructor" Einarsson. A miasma of *brennivín* encircled this multitalented man like a London fog. He had swollen, coddled-egg eyes. "*Komdu saell,*" he said. "So you have been talking to my chum, the murderer? Isn't he a fine person? His dead wife Birna went to school in Sweden. No Icelandic girl should ever go to school in Sweden. *Gud minn Godur!* The country is at least seven hundred years behind Iceland. But still I give Sweden a chance because of one man — Salomon Andrée — who flew to the North Pole in a balloon. *Skál!*"

"*Skál!* But Andrée never made it to the Pole. His balloon crashed. He died in Spitsbergen."

"So much for Sweden, then. But you" — here he gave me a fraternal rap on the chest that nearly knocked the wind out of me — "you will come to my house for supper tomorrow. You will meet my wife." He produced a photograph. "You will meet my three strong sons." He produced another photograph. "And best of all, you will meet my Iceland dogs." He produced three or four more photographs. "The Iceland dog is one of the rarest dogs in the world. They can talk to sheep in the sheep's own language. They win big prizes at dog shows. I am a breeder of Iceland dogs, also a seal hunter, taxi driver . . ."

"Hey, Yankee," another man broke in. "I give English a chance because of three writers — Shakespeare, Edgar Allan Poe, and Ray Bradbury."

"Tómas Gudmundsson is a better poet than Shakespeare," the man's woman friend said.

"Shakespeare has written of Iceland dogs," declared Einar, quoting *Henry V* in English: " 'Pish for thee, Iceland dog! thou prick-ear'd cur of Iceland.' "

"That is not very nice," said the other man, anger rising in his voice. He looked hard at me as though I were responsible for these unkind words.

"Tómas Gudmundsson would never write such rubbish," the woman said.

"*Skál!* Down with Shakespeare!"

"Down with America! Down with Ray Bradbury! Down with the Empire State Building! *Skál!*"

I was fast becoming the center of a literary controversy. It appeared that these people were going to make me pay for Ray Bradbury's and Shakespeare's crimes against the Icelandic state.

I excused myself and eased away gently, somehow ending up back with the murderer. He'd drunk not just the hair but the whole hairy rug of the dog that had bitten him the last time I saw him. He looked awful. "I wish I was back in prison," he said, and then collapsed peacefully on the floor.

By three A.M. any person who didn't already have a partner for the night was staggering around in search of one. It was like watching a ringful of punch-drunk boxers. The women showed slightly better equilibrium (lower center of gravity?) than the men, or at least they seemed less likely to fall down on the floor in pursuit of a mate. One woman lurched over to a man and butted her head against him. Another picked out a man and gave him what looked like a death-dealing figure-four leg lock. Others just careened around until they happened to crash into a member of the opposite sex.

This form of courtship has always impressed me. I find it a much more expedient method of finding a lover than making polite inquiries as to your candidate's interest in middle-late Rachmaninoff, quiet strolls on the beach, or noisy volcanic eruptions. Not only does it determine whether the person is still more or less conscious, but it also gives you some idea of his or her sexual potential. If he or she can respond in kind, with head butts, body blocks, and suchlike, fine; if not, it's probable that his or her bedroom performance will consist only of being sick or falling asleep.

Back outside, the sky was a deep cerulean blue. A lone white cloud rode directly overhead like a plucking of wool glued to a window. There was enough light for me to read a book or even — to use a popular old description of an Icelandic summer's night — pick the lice off my shirt. Despite the lateness of the hour, the ring of cars showed no sign of thinning out. I could still hear the bleat of Skoda horns, the irate Oriental falsetto of Datsuns and Toyotas, the outrage of Porsches, and the deep antiphonal bass of American Fords. All at once I heard someone shout my name. The shout came from a journalist friend of mine named Thorgeir who was driving something from an eastern bloc country powered by the smallest number of cylinders a car can have and still be alive. I got in, and we broke away from the circle. "*Öl er innri madur,*" Thorgeir greeted me. (Booze is truth.)

We ended up at the sluice of naturally hot water that runs from Reykjavik's reservoir tanks to the harbor. Called Thvottalaugur, Washing Springs, this rock-lined sluice is otherwise known as Thighs Canyon. Thorgeir and I stripped off our clothes and hunkered our thighs down among the thighs of elderly society matrons, teenage girls with breasts pointed gently skyward, hard-featured fishermen, taxi drivers, ship builders, heart surgeons, and rather overweight members of parliament. Thorgeir passed me a thermos of coffee laced with *brennivín* and someone else passed along a bottle of, I think, Stolichnaya. At some point I fell asleep. When I woke up, the Midnight Sun had been replaced by a sun of the more ordinary variety. I was sitting in Thighs Canyon, alone, naked, and parboiled.

Chapter 5

A TRIP TO HELL

ALL OF ICELAND has a Viking birthright, so I didn't feel I had to hunt down specific destinations where old-time settlers with names like Thorsteinn the Foul, Glum the Skull Splitter, and Gunnar the Unwashed had flung their chair pillars. A couple of these destinations — the Westmann Islands, for instance — had become too shamelessly prosperous or too heavily touristed (the Reagan-Gorbachev summit seems to have been, if nothing else, a public relations coup) for a connoisseur of last places like myself. Like my Viking trailblazers, I delighted in unvarnished rock and a paucity of people. I felt I could travel anywhere in Iceland, to uninhabited places, even *uninhabitable* places, and still be carrying on the spirit of my itinerary.

After nearly a week in Reykjavik, I felt the itch to move on, so I took a long rollicking bus trip across the interior of the island, along the blind mineral world of the North Atlantic Ridge, which brought me out of my urban and convalescent modes in a hurry. If Reykjavik is dedicated to the proposition that too many people is not enough, the interior is dedicated to the opposite proposition. *Nobody* lives there; nobody except a few summer-grazing sheep, arctic foxes, and minks escaped from fur farms. A desolation of lava fields, highland plateaus, tonsured mountains, and moraines, it seems to have been officially registered by the gods of the Tertiary as Off Limits to the human species. There are no signs indicating this constraint, but signs have always been forbidden in Iceland, as they make the desolation unsightly.

After we turned off the main coastal road, the pavement ended and a track began that had some of the most majestic washboarding ever to rattle my bones. Then the track ended and a kind of ash-and-cinder path began. But Icelandic buses are part horse and will crunch their way faithfully over any terrain, so this wasn't much of a problem. I even saw a few full-blooded horses in a pasture and they all turned to stare after us with, I thought, a certain envy. At last we crunched past Hekla, the culprit responsible for most of this blasted landscape, shaped not like the classic postcard volcano — after all, it erupted from a three-mile-long fissure rather than a single crater — but like the rotting hulk of an overturned boat. The only other traffic was large flatbed trucks hauling Hekla's pumice back to Reykjavik, where it would be mixed with concrete and used to insulate houses as well as to stonewash jeans.

Just north of Hekla I saw emerging from a gauze of mist Vatnajökull, the largest glacier outside of Greenland and Antarctica, looking like two hundred Matterhorns soldered together. Seeing it so suddenly was like having someone walk up to you on the street, tap you on the shoulder, and remind you of eternity. Eternity with a leak, however. For now we began to ford rivers from Vatnajökull's melt-off, together with rivers from Höfsjökull, a bare domed forehead of a glacier to our west. These rivers didn't flow, they hurtled themselves forward, fast and desperately hostile hurtles, with oceanic waves. But our driver ploughed right across them as if they were trickles from a faucet. Once we stalled in a particularly brisk stretch of white water, and it appeared we'd be upended and carried downstream a hundred miles to the sea, but our driver — who also, I might add, could change a tire with the dispatch of a short-order cook — rammed his foot down on the accelerator and off we raced to the gravelly bank on the other side. I made a mental note to bring a bus along on my next canoe trip; they aren't in the least deterred by contrary currents and they manage the portages with perfection.

Midday we came to Sprengisandur, a desert of rolling dunes at the top of a two-thousand-foot plateau, its black sand the result of rock crunching by Vatnajökull and Höfsjökull in their Ice Age incarnations. We started slithering along a barely visible track of sand a little like a roadway in the Empty Quarter of the Sahara, except that this sand was black and very, very wet. Then another Saharan motif appeared: a sandstorm. It whipped up out of nowhere and lashed the windshield with a relentless spatter of buckshot. I tried to look out the window, but couldn't see a thing except the blurry rim of a canyon along which we were slipping and sliding.

By now the trip was beginning to remind me of one of those infamous Latin American bus rides where the bus — actually a hodgepodge of cast-off tractor and automotive parts mounted on bald tires — bashes its way through mountains, swivels along precipitous gorges, straightens out hairpin curves, and generally avails itself of scenery to which no bus should have a right. Meanwhile the driver pulls off at every roadside shrine and leaves a bribe for the Virgin Mary; Mexican drivers leave iron washers in lieu of pesos, whereas Ecuadorians are more diligent and leave a certain number of sucres per wheel. But it wouldn't matter even if they left Her cassettes of salsa music. Sooner or later the bus will justify everybody's worst fears by plunging (Latin American buses never crash, they plunge) into a deep gorge, ravine, gulch, coulée, or canyon, the only survivor being a three-year-old child muffled by its mother's breast.

Fortunately Iceland has no roadside shrines to incite her bus drivers to acts of cockeyed bravado with their vehicles. It is a largely agnostic country where one priest I know praises venery from his pulpit and another rants and raves every Sunday against the American military in Keflavík. The only roadside shrines are the geological ones put there 50,000,000 years ago. Even so, I was beginning to think that a couple of stiff shots of *brennivín* might make the trip a little less dramatic. I was starting to go a bit stir-crazy, too. All those fierce rushing rivers, and I was denied the privilege of freezing my gonads wading across them! All that fine knife-edged basalt, and I wasn't able to savage my boots against it! We'd been traveling for eleven straight hours and it would be another three hours before we reached Akureyri, which wasn't even my destination. So after one of my fellow passengers got sick during a rough stretch of track (W. H. Auden, in *Letters from Iceland,* remarked that Icelanders were the only people he'd ever met capable of getting seasick on a bus), I asked the driver to liberate me from his jostling tube of metal. We were in the middle of nowhere, but I told him nowheres were my favorite places.

Once outside, I walked over to a little stream whose water had the not unpleasant taste of volcanic cinder. I was so glad to be off the bus that I wouldn't have minded if it had tasted of giardia. My feet were glad, too. They led me on a hike across the tundra to a glacial erratic that looked like battered busts of Roman emperors arranged in a semicircle. Map lichen went like a tapestry from boulder to boulder, scarcely missing a stitch. Here, in a silence so complete it brought a slight whistling to my ears, I pitched my tent. Then I occupied myself by splodging through deep blue-gray charcoal mud and gazing at a clump of arctic riverbeauties, a willow herb significantly more purple than anything else in the world. I could have stayed at this campsite for quite a few days. But the next day I was supposed to be in Reykjahlíd, on the shores of Lake Mývatn, where the geologist Gisli Saemundsson was expecting me. On the phone

Gisli and I had planned a little trip that would lead us into an even more bountiful nowhere.

Next morning I walked over six miles of tundra and bog to the Akureyri–Mývatn road and lifted my thumb in the time-honored gesture of the hitchhiker. The first time I did this in Iceland ten years ago, motorists hadn't the slightest idea what a raised thumb meant. Perhaps they thought I was saluting their courage in the face of potholes deep enough to swallow both them and their cars, for they invariably gave me a thumbs-up and drove on. Now times had changed for the hitchhiker, although the amount of traffic had not. After nearly two hours I got a lift to Reykjahlíd with a Land Rover full of Japanese scientists whose work presumably had something to do with glacial rivers at flood stage, because their leader told me: "We have lost one of our group. He drowned in river. We shall miss him." Then he grinned, it seemed to me, unnecessarily.

Gisli had said to look for a plain white bungalow with a flower-decorated whale vertebra on the front lawn. Yet in Reykjahlíd, a village of some two hundred people, nearly everyone seemed to own plain white bungalows with whale vertebrae on their front lawns. It was uncanny. I wondered how frequently it happened that a local ended up in the wrong plain white bungalow by mistake, set up housekeeping with the wrong spouse, and lived out his years as the wrong person. This never could have happened with the scruffy old turf-and-stone houses of fifty and sixty years ago, with sheep grazing on the thatch of their roofs; a countryman always knows his own sheep.

Finally a farmer directed me to the right house with the right vertebra on its lawn. Gisli was at home, and so was his large black mastiff, which immediately made a lunge for my throat. He patted the dog affectionately on the withers, saying, "Snati does not care for strangers." Then we retired to a living room whose walls were filled with books, primarily the Sagas and geological manuals, though I did notice a copy of *Likamsraekt med Jane Fonda* shelved between *Egil's Saga* and *Grettir's Saga.* His wife, Gisli said, swore by Jane Fonda; he himself preferred father Henry, and thought *The Grapes of Wrath* the best movie ever made, except for, maybe, *Stagecoach,* but he admitted he hadn't seen that many movies. In fact, he'd seen almost no movies until two years ago, when the local food market replaced some of its produce with videos.

To broach the purpose of your visit before the fifth or sixth cup of coffee is, to an Icelander, unspeakably rude. So we sat around drinking cup after cup of coffee — that the samovar wasn't invented on this volcanic island seems a curious oversight — until after the fifth cup? the sixth? the tenth? Gisli announced:

"Let's go to Hell!"

A word about Hell.

In the winter of 1875 a cold rain fell for weeks on the west coast of Norway. Then on Easter Monday, March 29, the keeper of the Ona lighthouse included this ominous note in his weather report: "Tonight, between eight and ten o'clock, fine, grayish sand fell with the rain, forming a layer two inches thick which stuck to the window panes and house walls." A day or two later sand began falling over Sweden, burning the eyes of anyone who ventured out of doors. In Södermalm a gardener noticed his plants were dying because the sand blocked the sunlight from his greenhouse. In Riga, Latvia, the local press wondered whether the sand might not be some sort of obscure fallout from German industry, blown eastward. The sand traveled to St. Petersburg and Moscow, terrifying the Russians, who thought their snow had changed color. Later that spring Irish potatoes fell prey to a mysterious blight, thought to be the same one that had struck them so disastrously in the 1840s.

As scientists and citizens cast about for a suitable explanation, several leading luminaries proposed that the sand was extraterrestrial in origin. The Swedish explorer Baron Nils Adolf Nordenskjöld, first man to navigate the Northeast Passage, called it "cosmic dust" and thought it came from the Moon or maybe Jupiter. Such dust, he surmised, had been falling on our planet for countless millennia and possibly even made up the very ground we walked on. Others aimed their telescopes at Outer Space to see if there mightn't be a puncture somewhere . . .

As it happened, the sand was volcanic ash from an equally remote but not cosmic place, the no-man-fathomed Askja region of northeastern Iceland. Specifically, it came from a new crater Icelanders themselves called Vití — Hell. There were already as many Hells on their island as Shady Groves in Illinois, but this one had such a diabolical effect on agriculture that it sent the first wave ever of Icelandic emigrants scrambling for the New World. The crater that made them scramble was scarcely bigger than a putting green.

We went to Hell in a four-wheel-drive Isuzu Marauder, a combination Land Rover, tank, and ballerina, its chassis so high off the ground that it seemed to be standing on tiptoe. It was the ideal vehicle for fording glacial rivers, subduing primeval and/or snowy roads, and pirouetting in the lava. Gisli said the only mishap he's ever had with his Marauder came last summer when a French tourist skidded on the midge corpses (Mývatn means Midge Lake) littering the road and slammed into him.

"The French are terrible drivers," I said, stating a commonplace.

"You should see them trying to drive on dead insects . . ."

First we made a detour east to Mödrudalur, a hundred years ago the most remote farm in the country, but now more or less connected to the rest of Iceland by the main ring road.

Mödrudalur squatted on a broad highland plateau whose sintery ground was pale dun barely relieved by little clumps of grass. Wind devils whipped up swirls of sand and swept them along for several hundred yards until they spanked against ash heaps and fell apart. At any moment I expected to see a procession of Bedouins emerge from one of the wind devils or hear a muezzin's monotonous nasal *La ilah ill' Allah* issue from the farm's miniature church (the plans for which, according to Gisli, were drawn up by farmer Jón Stefansson in six minutes).

An Englishman named John Stanley visited Mödrudalur in the nineteenth century and noted that its occupants could recite *Paradise Lost* chapter and verse in the Icelandic translation, but had no idea what the apple was that Milton kept referring to. Just as a traveler to Africa might have astonished the natives by taking out his glass eye if he were fortunate enough to be so equipped, Stanley produced an apple and the natives of Mödrudalur were mightily astonished. They kept this holy relic for many years, until it was virtual dust.

Heading west, we came to the crater of Hrossaborg — Horse Crag — and traded the main road for a track of sandy grooves leading into the interior. The withered plain was punctuated only by small hummocks of sand and ash. Brave clumps of chickweed and campion huddled together like survivors of an antibotanical blitzkrieg. We began to run parallel to the Jökulsá á Fjollum, a dark and murky glacial river littered with more rubbish, albeit geological rubbish, than the Rhine.

Now the track got worse. Gisli did not hesitate to call it the worst thoroughfare on an island of bad thoroughfares. High praise, indeed. It had been cut from a lava field called Grafarlönd using only hand tools — picks, shovels, and chisels — in the early 1960s. The track kept switching back on itself like a pretzel, as if it resented mechanized vehicles and resisted them by not going anywhere. Gisli utilized every erg of his four-wheel drive. The Marauder turned into a hesitant burrowing mole, abruptly stopping every few feet, backing up, and then proceeding again over a catacomb of tangled torsos and crushed basaltic bones still spiky with life. Softer springs, and we'd have suffered the perils of impalement.

Gisli told me this story:

At the turn of the century there was a man who wanted to get married but had no money, so he signed on with a Norwegian whaling vessel. After a profitable season, the man returned home and tied the knot of marriage. Three or four years later he was hard up again and signed on with another whaler. One day they took an enormous blue whale off Grytviken, South Georgia, and the man was assisting in flensing this whale when he tripped and fell onto the tongue. The tongue of a blue whale is six feet thick and extremely pulpy. The man sank into it and suffocated to death.

This man was Gisli's grandfather.

"I grew up in Siglufjördur, a Klondike for herring. Every boy in Siglufjördur went to sea except my brother and me. We became geologists. I think we chose a profession where we would be able to work with hard material, very hard material . . ."

After the Marauder made a dash across the river Lindaá, the track seemed to become a little less resentful of our intrusion. Every once in a while I even saw clumps of angelica, an edible flowering plant that looks exactly like spotted cowbane, one of the world's *least* edible plants (one mouthful can kill you). Then I saw Herdubreid slumbering in the lava directly ahead of us. It was shaped like a heroic birthday cake, with a lone candle for a summit and icing in the form of snow dripping down its numerous couloirs. Most mountains seem to lose their personalities and become amorphous hulks the closer you get to them, but the closer we got to Herdubreid, the handsomer it appeared.

It was the mountain's ash and tephra, Gisli said, that made climbing it such an arduous task. Two steps forward usually resulted in three skids backward. Later I learned that the first mountaineer to climb Herdubreid was an American named W. L. Howard. Howard made his ascent in 1881 with the help of a kite fastened to an anchor. He'd throw the anchor onto the rocks above his head and then pull himself up by means of a rope, the kite bearing him aloft whenever he reached an impossible rock face. I could imagine the difficulty of constructing a kite, in the interior of Iceland, of sufficient size to lift an anchor along with a brawny mountaineer. Nevertheless, Yankee ingenuity — inspired, perhaps, by Benjamin Franklin's noble experiments with kites — seems to have prevailed, and Howard managed to turn a six- or seven-hour climb into a thirty-eight-hour climb.

We got out at Herdubreidarlindur, an oasis of sorts in the lava desert. Along a fierce little brook there grew pink thyme, arctic cotton, and the ever-present angelica, some of which we ate *au naturel,* though Gisli informed me that it was best with the raw yolk of a duck egg poured into its hollow stem. Then we walked to a rubble of stones that had once been a shelter (something like a Scottish bothy) lived in by two of the only inhabitants of the interior wastes, Fjalla-Eyvindur and Fjalla-Bensi, respectively. Eyvindur was a sheep thief and Bensi was a shepherd, but the mountains that gave them their nickname, Fjalla, hardly observe such paltry distinctions.

Eyvindur Jónsson was born in 1714 and enjoyed a quite normal childhood except for his constant desire to steal. He started out with small things, like household utensils, and then worked his way up to livestock. Finally, after a large quantity of stolen sheep were found in his possession, the Althing proclaimed him an outlaw. Now he took to the mountains and pursued his trade in earnest, stealing sheep morning, noon, and night from rich and poor alike. No account of his career ever mentions personal gain or a violent craving for mutton. Eyvindur simply stole sheep as some other talented lad might have written verse. Indeed, one of the few descriptions of him warns his potential victims that "he is frequently heard humming strophes of *rimur* or old songs." He would arrive at a farm — incognito, of course — and first win the farmer's trust. *Then* he'd make off with the man's sheep, as if it were wrong just to descend on the farm and steal the sheep. When it came time for him to marry, he made certain the woman of his heart, a young widow named Halla Jónsdottir, shared his interest in sheep thievery. If a farmer or sheriff got too hot on their trail, Eyvindur and Halla would disappear into the uncharted interior. Eyvindur himself was no faster on his feet than the next man, but he had a strategy that enabled him to escape even his swiftest pursuers — he turned cartwheels. An entrancing vision this is: a lone outlaw cartwheeling wildly over lava and sand, as the forces of law and order lag ever farther behind him.

"Eyvindur hid here," Gisli said, "during the winter of 1774–75. He had a dead horse for a roof. He ate the meat from this horse and by spring the sun was shining through its bones. A year later he was pardoned. Twenty years he'd been an outlaw. The government won't let you be an outlaw longer than that."

I looked at this heap of stones and tried to imagine someone living there. In its prime it would not have measured more than five feet by three feet, hardly enough room to turn a single enervated cartwheel in. But Iceland seems to prepare a person for the stoniest of conditions and fosters in him the most creative of survival tactics: if a dead horse is your roof, at least it keeps out the weather.

Whereas Eyvindur stole sheep, Bensi retrieved them. He was a scruffy, snuff-taking farm laborer whose job it was to find missing sheep. Once he struggled for days to bring back a single sheep to Grimstad farm, but then he lost it. This legendary effort — only a slightly dotty person like Bensi would have been so worried about a lone sheep — provided Gunnar Gunnarson with the idea for his 1944 novella *The Good Shepherd,* which is about a snuff-taking shepherd's failure to retrieve a sheep during a Christmas Eve blizzard. According to Gisli, Ernest Hemingway cribbed *The Good Shepherd* for his novella *The Old Man and the Sea* ("Every Icelander knows how Hemingway stole from Gunnar Gunnarson"), turning Bensi into the old Cuban fisherman, the sheep into the tenacious marlin, and the Christmas Eve blizzard — this I found a little improbable — into the Caribbean Sea.

"I've often wondered why he committed suicide," Gisli said.

"Who? Bensi?"

"*Nej,* Hemingway. Do you suppose he felt guilty about stealing from Gunnar Gunnarson?"

We continued westward parallel to Herdubreidartögl, Herdubreid's Horsetails, a long muscular scarp of palagonite which looked less like horsetails than the recumbent body of a weight lifter. On either side of the track, tortured Modigliani bodies of black lava rose up, their gaping mouths seeming to try to tell us something: *Et in Arcadia ego,* you complacent wheeled bipeds. Now I began to see hoary-headed Medusas and mutant reptiles and huge buzzards and malevolent gnomes crouched on battlements of broken-down rock, all waiting to pounce. Once I saw a dried stalk supporting a frail flower; it seemed like an evolutionary mistake.

In the chill and quietude nothing stirred, not even an insect. The whole effect was like passing through a bombed-out city. I thought of Dresden. Even more I thought of Hiroshima and Nagasaki, for Icelanders have a joke that a nuclear attack would be redundant in their country.

"We are now within the Ódádahraun," announced Gisli in a tone of reverence, as if we'd just entered a church rather than the biggest, emptiest lava field in the world.

The Ódádahraun, Lava of Evil Deeds, gets its name from the outlaws who supposedly followed Eyvindur's example and holed up here in the nineteenth century. But nothing has ever been found in this 2,000-square-mile tract of science fiction to indicate occupancy by outlaws. Occupancy by astronauts is a different matter. In 1967 NASA authorities sent Neil Armstrong, William Anders, Alan Bean, and other members of the Apollo 11 mission to make a few preliminary steps for mankind among the demented shapes of the Ódádahraun. It was a perfect stand-in for the Moon; other lavas in other parts of the globe have been defiled by vegetation or tampered with in some way and thus are not sufficiently lunar. The Apollo astronauts arrived, hopped around a bit, collected rock samples, identified rock samples, threw away rock samples, and played American football in preparation for their epochal trip two years later.

"We were pleased to host your astronauts," Gisli said.

"They're not *my* astronauts," I replied a little peevishly. For I am not too kindly disposed toward these rocketeering holidays in space. Whatever can't be reached with a good pair of hiking boots or a mercifully slow boat ought to be left alone to revel in its own devices. The Apollo crew had already visited the Ódádahraun: was it really necessary for them to pay a flying visit to our lunar neighbor, too?

This, I realize, betrays a somewhat limited, if not wholeheartedly narrow, point of view. Yet I've tried to be openminded. Tried to regard the Moon, Jupiter, Pluto, Betelgeuse, the Andromeda galaxy et al. as merely more islands in the wilderness of Creation. Tried to defer to the adventurous spirit of the Vikings when I contemplate (for instance) the Viking space mission. But it doesn't seem to work. I remain hopelessly earthbound, a prisoner of my own parochial outlook, my parochial feet. Like everyone else, I've gazed with awe at twinkling stars in the clear nighttime sky, but my brows somehow refuse to knit over the prospect of extraterrestrial tenants and their possible IQs. All I hope and pray for is that these stars can hold on to their wondrous twinkle despite the achievements of Terminal Man at his computer.

Perhaps I'm just reactionary. Backward. Thick-witted. Unscientific. I've even read one or two books about the Apollo Moon mission, yet I confess I still can't figure out what Armstrong was doing up there on July 20, 1969. You can't play the trumpet in a vacuum . . .

By early evening we reached the hut at Drekagil, Dragon's Ravine, which marked the end of our trip. The end of the automotive part of it, that is. We had to traverse the remaining four miles on foot.

Soon we were slogging through a snowfield under which rivers of melt-off sang Siren tunes, an invitation to the heedless walker to take a wrong step on the snowy scurf and plunge in. We clattered over a stretch of pumice, circumnavigated pyramids of ash, slogged through more snow, sank into sand, and at last stood on a ridge above the primal amphitheater of Öskjuvatn, a deep pear-shaped crater lake two miles in diameter. A few stray slabs of ice were floating in the water, but otherwise its surface was imperturbable.

We stood there, not saying a word. What words can do justice to such inanimate, inhuman, unfathomable loveliness? Across the lake a mass of quartz gleamed white and lonely as Lot's wife. Noctilucent clouds drifted by like plump golden dirigibles. I took in some of the purest oxygen I've ever tasted, took in more and more of it, and slowly a sense of ancient calm eased its way through my body's defenses. More oxygen; more clouds; more hermetic lake. I was becoming One, no, I *wasn't* becoming One with the world; that was a physical impossibility. But I was becoming Two with it, and each of us was on the very best terms with the other while observing our respective identities. An ideal relationship, I thought.

Gisli broke the silence with this story:

In the early summer of 1907, an expedition of three Germans left Mývatn for the Askja region. The group consisted of the renowned geologist Dr. Walter von Knebel, a young geologist named Hans Spethmann, and the painter Max Rudloff. They arrived at Öskjuvatn with several Icelandic guides and a boat made of canvas attached to a brass frame. After the guides left, the Germans settled in on the shore of the lake. Their tents were the first ever to be pitched there. The two geologists collected a number of remarkable rock samples from the 1875 eruption, and Rudloff made sketches of a landscape he never would have thought possible outside a grotesque storybook.

On July 10 von Knebel and Rudloff set out to explore the lake while Spethmann trekked into the mountains in search of more rock samples. Spethmann came back to the camp early that evening; the others did not, nor could Spethmann see their boat floating anywhere on the lake. Five days later they still hadn't appeared. When one of the guides returned to the camp, Spethmann was wandering around half-crazed and incoherent, unable to explain what had happened. It was soon rumored that he had done away with his two companions because he was in love with von Knebel's fiancée, Viktorina vom Grumbkow. Later Fräulein vom Grumbkow mounted an expedition to Öskjuvatn, vowing that she'd get to the bottom of the mystery. No doubt this would have meant getting to the bottom of the lake, for which she would have needed a bathysphere. After a fruitless search, she tossed a funeral wreath into the lake and went back to Germany, where she married a vulcanologist, Dr. Hans Reck. At this point the rumor arose that Spethmann hadn't been interested in her at all, but that Reck had offered him a tidy sum to dispose of the competition. As for von Knebel and Rudloff, nothing has ever turned up except one oar, one of Rudloff's gloves, and a paintbrush.

Gisli showed me the cairn Fräulein von Grumbkow had erected to the memory of her fiancé. Under the capstone was a rusty tin box with a logbook whose previous entry, in English, read: "No Nukes!"

We moved on, now traversing ground both tangerine and somber indigo in color. Soon we came to a small crater girded by a pumice cone and separated from Öskjuvatn by a thin rock wall. From this crater twisted and gyrated phantoms of windblown steam. Unlike Öskjuvatn, it was filled with *living* water — water with the complexion of green coffee and the aroma of terminally rotten eggs. So this is Hell, I thought. The same infernal realm that had savaged Irish potatoes, scattered Icelanders to the United States and Canada, and broken through the Russian winter with salvos of ash. It looked more like a dirty bathtub.

I followed Gisli down a steep clay embankment, greasy as Vaseline, into the crater itself. At one point he slipped and fell rudely on an abrupt outcropping of gray quartz. *"Ja madur, mikid helvití!"* he exclaimed. (Lousy damned diabolical place!) An expression of disfavor, for an Icelander, equal to any his thousand-year-old Viking tongue has to offer.

Actually Gisli had great affection for this place. As farmers around Lake Mývatn swear by the magical powers of horse shit, especially when smeared all over their hands, so he swore by the enchanted waters of Hell, which brought him good luck, especially when he swam in them. Eight years ago Hell had landed him a job with the Krafla geothermal project, and just last year it had gotten him a pretty young wife (though she wouldn't sleep with him for a week after he swam here, due to the smell).

When we reached the bottom, we took off our clothes and waded in. Once again I was in hot water — hot, sulphurous water that now had the odor of particularly aggressive onions along with its rotten eggs.

Within a few minutes fat flakes of snow were sifting down into the crater. The heat liquefied them before they reached us. But soon the snow came faster, and I found myself in the unlikely position of doing the sidestroke in an Arctic blizzard and being hot and cold simultaneously. Not only that, but in the midst of this blizzard the sun came out and targeted a patch of green water with one of its nocturnal beams. So many mutually exclusive things happen in Iceland that at this instant I wondered why they didn't cancel each other out and leave the island drearily normal.

That night, at the Drekagil hut, Gisli fell asleep right away and soon was snoring like a Mahler symphony. I couldn't sleep at all. It was one-thirty A.M., and muzzles of light were still licking the window. I wished I had two blankets: one to cover the window, the other to cover Gisli's face and quiet down his concertizing.

Thus I put on my boots and headed back outside. All of creation was a mild Arctic blue, the only sound was the drip-drip-dripping of melting snow, and low stratus clouds hung peacefully in the sky. A perfect night for a stroll. Where, I wasn't sure, but I knew I wouldn't get lost because I could always check my bearings against Askja's castellated ridges and the purple wedding cake of Herdubreid north-northwest in the distance.

I climbed up the butte behind the hut and walked along a plateau where the first tentative growth of moss covered the lava like verdigris. I hopped over crevices like an astronaut and walked through lava grottoes so flawlessly grottolike they could have been hand-sculpted. Just outside one grotto I found, not Rudloff's other glove, but a Prince Polo chocolate wrapper, which indicated to me that an Icelander rather than a German had once passed this way. Prince Polo is the island's most popular chocolate bar, eaten by everyone addictively and in such quantity that it's been said that only the sweet tooth of Iceland keeps the Prince Polo factory in Lodz, Poland, from being replaced by a tractor parts factory.

Suddenly a curtain of fog came down like a concentrate of pea soup and oyster stew. All I could see was my proverbial hand in front of my face. I certainly couldn't see the two landmarks I had noted earlier as compass points, so I decided just to sit down and wait out the fog. Fifteen minutes passed. Half an hour passed. The fog showed no sign of lifting, even though the wind had shifted to the blow-away direction, east. I could be fogbound at this spot for days, I thought; back in 1838 the first known visitor to Askja, a surveyor named Björn Gunnlaugsson, *had* been fogbound for days. So I decided to retrace my route to Dregagil by following the foggy silhouettes of the lava formations I'd seen earlier as well as my own footprints in the snow. A mistake. Lava formations look different from different angles, and much of the ground I'd been walking on was gravel and pumice, not snow. However refined your ability as a tracker, it's not easy to detect footprints on gravel.

I got lost.

I had been lost before — many times, in fact — but at least I'd had a compass and a topo map and, most important, I could see the territory into which my misdirection was leading me. Here all I could see was a billowing silkworm cocoon and a few dark phantasmal shapes, maybe lava, maybe a bestiary of strange animals. I started to walk a bit faster, jumping over slag heaps and wading through knee-high drifts of snow. I cut my hands and scraped my legs on sharp poniards of lava, one of which drew a thin red line of blood through my hiking trousers.

I seemed to be heading in the opposite direction from Dregagil, but when I doubled back, *that* turned out to be the opposite direction, too. Once my feet shot out from under me and I went head over heels into a lava hole. Luckily it was a small lava hole, as the big ones can coax a person to a rather lengthy stay. A couple of years ago a set of bones was found in a hole near Kalmanstunga that still had pieces of clothing from Settlement Days clinging protectively to them.

I climbed from the hole and continued to carve a sort of passage through the fog. Another shift in the wind brought the sulphurous smell of Vití to my nostrils, which made me think I couldn't be too far off course. But then I remembered that I'd smelled Vití even at the Drekagil hut, so I might not be that close, after all. Then where was I? Maybe Cambodia, because I nearly smashed into something that looked like the Temple of Angkor Wat being taken over by jungle. Or maybe Paris, for now I saw the Arc de Triomphe etched distantly in the fog. *Paris!* A certain part of me would have loved to be there right now, ordering an *apéritif* and a *cordon bleu* meal from an unbearable waiter. But another part of me preferred to be here in this mysterious nowhere, colliding with lava formations. (What better way to meet them?) To hell with Paris.

Then it happened: All those dark phantasmal shapes turned into a bestiary of animals, becoming strangely, exultantly alive. This wasn't an adrenalin surge increasing a panicky mind's level of awareness. They *were* alive — trotting, promenading, and rejoicing in the fog as if it were the one and only catalyst in their otherwise stationary existence. Birds, reptiles, and great shaggy mammalian beings with heads held pridefully aloft. Gnomes and trolls who had shaken off their supernatural malaise like butterflies shaking off their cocoons. All of them were trying to tell me something: namely, that I was an intruder in their laval dust and the sooner I got out, the better it would rest with them. Several even indicated the route I should take:

"Head northeast over that tenebrous ridge, Stoneless One."

Their directions turned out to be surprisingly good. I trudged over the ridge and walked for another hour along a swiftly moving glacial river. At last I saw the Drekagil hut down a loping incline only a quarter of a mile away. By the time I reached it Gisli was already drinking his third? fifth? seventh? cup of morning coffee.

A lucky thing *had* happened, the result, maybe, of yesterday's swim: I was found. Still better, I'd had the privilege of being hopelessly lost in the outer neighborhood of Hell.

Chapter 6

DOG-DAY REVOLUTIONARY

AFTER MY TRIP with Gisli, I hitched a lift north to the port of Akureyri, where I would catch the ferry to Grimsey. While I was waiting at the pier, I looked through a book I'd picked up in a local bookstore about the Danish soldier of fortune Jorgen Jorgensen. Icelanders refer to Jorgensen as their Dog-Day King because he came to power during the so-called dog days of one very remarkable summer in the year 1809. Actually, the island can't claim real dog days — days when the Dog Star, Sirius, shines just before sunrise and just after sunset — since the Midnight Sun runs interference with lighting conditions. But then Jorgen Jorgensen wasn't a real king, either.

A dockhand saw me sitting there and peered down to find out what I was reading. "Ah, Jörundur," he said, using the Icelandic sobriquet for Jorgensen. "*Hamingjan góda!* He was a man with a very strange story . . ."

Here is that strange story:

It is the year 1780, only two years since Captain Cook has discovered Hawaii and still three years before the end of the American War for Independence. In Copenhagen, Denmark, the wife of the royal watchmaker gives birth to a son named Jorgen. From earliest youth, perhaps because his father is so obsessed with time (Mr. Jorgensen has written books in Danish, French, and German about watches and chronometers), Jorgen seems to delight in disorder. He likes to see things broken, violated, ravaged — especially governments. Especially the government to whose court his father pays obeisance. Thus the single most satisfying event of his youth is standing with a crowd of onlookers and watching the king's palace in Copenhagen burn down. At the time of this happy event, he is fully nine years old.

But a person can hardly expect to make a career out of hating monarchy. Jorgen decides to be a sailor. At age twenty he journeys to England, land of sailors, and is wandering the docks at Southampton when he is impressed on board a Royal Navy vessel. Soon he is cannonading French frigates under the adopted name of John Johnson. For his cannonading abilities, Ordinary Seaman Johnson becomes Second Mate Johnson and joins the *Lady Nelson,* under assignment to explore the distant coasts of Van Diemen's Land, now Tasmania. In 1804 he arrives in Van Diemen's Land and helps lay the foundation stones for the capital, Hobart Town — a place, he remarks to a friend, he'd like to see again some day. Fateful words!

Next Jorgen is Chief Mate Johnson on a whaler, the *Alexander,* sailing from New Zealand to London with a rich cargo of sperm oil. Rounding Cape Horn, the *Alexander* meets up with one of those once-in-a-lifetime storms that blows it 3,000 miles off course, to Tahiti. Tahiti, Jorgen finds, is even more congenial than Van Diemen's Land. Here he dallies with native girls whose mothers had given such a full-bodied welcome to the original *Bounty* mutineers. He meets a missionary (missionaries, he notes, "are selected from the dregs") who says that in all his years in Tahiti he has made only one convert, a retarded fourteen-year-old half-caste. He also meets King Pomare the Second, six foot five and bulky in proportion, who welcomes white visitors with "Master Christ very good, very fine fellow. Me love Christ like my own brother. Please give me one glass brandy . . ."

The *Alexander* at last returns to England with two Tahitians, who become instant ethnological specimens named Jack and Dick. Jorgen sails on to Denmark, only to find his country preparing for war with England (Denmark was an ally of France under Napoleon). As a sailor and navigator, he is given command of a Danish privateer, the *Admiral Juul,* with orders to blow up British brigs. Before he can blow up a single brig, however, he is taken prisoner and landed at Yarmouth, where he is set free on parole. Right away Jorgen hops his parole by heading up to London and purchasing himself a new wardrobe.

In London, Jorgen decides to look up his former charges Jack and Dick, who are currently being studied by the great naturalist Sir Joseph Banks. Jorgen and Sir Joseph take to each other almost immediately. Somehow the conversation shifts from Tahiti to Iceland, a country Sir Joseph visited thirty years earlier. Since that time the Danish administrators of the island ("You will pardon me, Mr. Johnson, if I speak harshly of your countrymen") have dissolved the Althing, symbol of a thousand years of democracy, and relegated its function to a local lawyer. And now, adds Sir Joseph, there is even more cause for concern, for Iceland is wholly dependent on Denmark for her imports and Denmark is presently engaged with the European war. Wretched people! Lately the Icelanders have been obliged to distill seawater for their salt. Sir Joseph has even heard stories about how some of them must perforce eat their own lice, like the indigenes of Kamchatka.

Jorgen has always felt a sympathy for the wretched of the earth. It isn't long before he has convinced a wealthy merchant, Mr. Samuel Phelps, to provision a boat, the *Clarence,* with salt, barley meal, and potatoes. Then in January 1809, he sails with Mr. Phelps to Iceland in the hope he can relieve some of the island's wretchedness. In Reykjavik he finds a motley of his countrymen in the town's lone tavern, smoking their clay pipes and drinking ale, a corrupt modern race indifferent to the ancient race over which they've gained ascendancy. Jorgen locates the factor and asks him for permission to trade, but is denied that permission because Iceland is a Danish monopoly and has been ever since the year 1381. Well do I know that, replies Jorgen, but I am a Dane, so trading with me would not be illegal. Permission again denied. The factor, slightly in his cups, laughs: "Here is a Dane arriving on an armed English vessel provided with letters of marque!" Jorgen sails back to England all the more determined to break the Danes, whom he's always hated with the passion of a native son and now hates even more.

In June 1809 another armed vessel arrives in Reykjavik. She is the barque *Margaret and Ann,* under the command of Captain John Liston, with Mr. Phelps, the young botanist William Jackson Hooker, and Jorgen himself on board. While Hooker sets off in search of flora, the others seize Count Trampe, the Danish governor, and march him back to the boat under armed escort. Captain Liston confiscates the island's entire arsenal — twenty-five antique fowling pieces. As for Jorgen, he walks up to Government House and blithely proclaims himself head of state: "We, Jorgen Jorgensen, His Excellency the Protector of Iceland, Commander-in-Chief by Sea and Land." He opens up Reykjavik jail and drafts the prisoners, primarily drunks, as his personal bodyguard. The next day, June 26, he issues a proclamation stating that the days of Danish rule in Iceland are over. Terminated. Gone forever. Henceforth his subjects will be free to trade with whomever they please. His Excellency himself designs a new Icelandic flag showing three dried codfish against a field of blue.

In his first and last public speech, Jorgen makes it clear that his coup d'état is a revolution and not a British occupation of Iceland. This seems to surprise a number of his subjects, who had always thought revolutions required lopped heads, like the French one, or a healthy body count, like the American variety. In this revolution, though, only twelve men have been employed, not a drop of blood shed, and not a single shot fired (except by one of the prisoners, who got drunk and accidentally shot himself in the leg).

Now His Excellency mounts his horse and travels across the island to observe the daily life of his subjects. His journey takes him past salmon rivers and bilberry-covered heaths, over lava fields and into the snowy realm of mountains from which glaciers tumble like frozen cascades. Along the way he stays in native farmhouses and samples native cuisine (pickled ram's testicles and horse steaks). In these farmhouses the farmer, his wife, their children, and the hired help all sleep in the same bed, the better to keep themselves warm at night. Not infrequently Jorgen joins them in these same beds. From now on there are rumors of little Excellencies playing in the ashes of native hearths.

Seldom has a less likely sovereign graced the corridors of power. Jorgen thinks taxes are boring, believes in paying teachers well and judges badly, and thinks redistribution of wealth a capital idea. Of course, the gentry whose wealth His Excellency is busily redistributing think much less of the idea; one man, Judge Isleif Einarsson, ends up in jail when his plan for a counterrevolution is divulged. A magistrate from the Skaftáfell district named Jón Gudmundsson declares that if Jorgen and his entourage ever appear east of the Jökulsá á Fjollum they will be treated as common outlaws. The grand old man of Icelandic politics, Magnús Stephensen, despises Jorgen, the revolution, and the codfish flag alike. He even invites His Excellency to dinner with the express intention of making his displeasure known, though his advocacy of Danish rule loses its edge somewhat when he sits down at table and his chair collapses under him.

On the evening of August 14, events take a decided turn for the worse. This, indeed, is a crucial date in Jorgen's career. Henceforth he will be bucking an uphill course against the dictates of power and his own self-destruction impulses, and as he bucks that course, his idealism by slow degrees will fade away. On this evening the British man-of-war *Talbot* sails into Reykjavik harbor under the command of the Honorable Captain Alexander (Paddy) Jones. Captain Jones is an Irish snob with the prejudices of an English snob, and Jorgen's butterfly escapade of a revolution is plainly not to his taste. This taste he can back up, if necessary, with a crew of five hundred men and numerous cannon. His Excellency, he concludes, has played the traitor to both England and Denmark; he has acted in open rebellion against both countries, but especially against Denmark, which Captain Jones apparently feels Britain cannot afford to irritate even though the two countries are fighting each other tooth and nail. Soon Captain Jones is lowering the codfish flag, obviously a joke. Not a joke? Well, so much the worse for Mr. Johnson.

On August 22 His Excellency Jorgen Jorgensen, a.k.a. John Johnson, is forced to abdicate. His army of eight is disbanded, and all Danish property and valuables are redistributed to their former owners. The dog days — and their king — have fled with the summer itself. Iceland digs in for the long winter, restored to its former capacity as a Danish colony, which it will remain until 1944.

Back in England the former Protector of Iceland is sent to Toothill Fields Prison, where he falls in with a checkered crew of cardsharks and gamblers who educate him in the intricacies of their trade (or has he been a gambler, of a different sort, all along?). After four weeks he is removed to a pestilential prison ship, the *Bahama,* where he takes up his pen and writes a defense of the Icelandic Revolution in the form of an autobiographical novel, *The Adventures of Thomas Walters.* Then he writes another novel, *Shandaria,* which combines Iceland and Tahiti into a single Utopia vaguely described as situated in Asia just beyond the dominion of the Great Mogul — a Utopia with no poverty, no prisons, no drunkenness, no lawyers, no doctors, no policemen, and no communication with foreigners. In short, no evils whatsoever. *Shandaria* he dedicates to his old friend, the botanist Hooker.

Hooker himself at last secures the Protector's release from the *Bahama.* Jorgen heads directly for the gaming tables where he promptly loses everything and is thrown into Fleet Prison, the debtor's prison. Prisons seem to bring out his literary side. In Fleet Prison he meets a fellow inmate recently returned from Afghanistan and cribs a memoir of that country based loosely on his friend's exploits. The memoir somehow falls into the hands of the foreign secretary, Lord Castlereagh, who reads it with interest, for its author seems to be describing precisely the route the Russian czar would take if he elected to invade British India. This fellow Johnson, Castlereagh realizes, is exceptionally clever. He engages the former Protector of Iceland as an agent in the British Secret Service and sends him to the Continent to spy on France.

The best vantage point for spying on France, Jorgen decides, is the gaming tables of Paris, to which he repairs upon his arrival. He promptly loses his shirt, as well as his jacket, gaiters, and shoes. A bleak December morning in 1814 finds him quitting the French capital on foot, penniless and nearly naked. Yet before the end of the day he's posing as an Irish pilgrim en route to the Holy Land and receiving much-needed alms ("I thank ye, and Our Lord Jesus Christ thanks ye, your honor") from devout country folk. He heads east into Germany and gets an audience with Goethe, introducing himself as the exiled king of Iceland, now sadly reduced in circumstances. Likewise he meets Prince Püchler von Muskau, who takes him up in a hydrogen-filled balloon. Someday, he tells the prince excitedly, everybody will travel by air. That is not likely, my friend, replies the prince, who cannot imagine anyone but aristocrats like himself and His Excellency taking to the air. Meanwhile Jorgen is scribbling away at a top-secret report for Lord Castlereagh which blames Napoleon for every penniless beggar and every ruined gambler in Europe. This does not seem likely, either.

Back in England the former Excellency is arrested for trying to pawn his landlady's furniture. He lands in Newgate Prison, whereupon he begins work on a history of Madagascar, plundering information from another prisoner, a former captain of a French slaver. This time no one in the foreign service seems interested in the manuscript, so he begins a history of the Christian faith, which seems a more marketable subject than Madagascar. One day while he's working away at this latter tome, he learns that the Court of Appeals has granted him a pardon on condition that he leave Britain and never return. *Never return!* What, he wonders, has he ever done to justify being banished like this? But he's delighted with the pardon, which he celebrates by hitting the gaming tables again and once again losing everything. Back in Newgate Prison, this time he's sentenced to be hung. "In a rather dreadful scrape," he writes his friend Hooker, who by now must have wished he'd never met His Excellency. But once again Hooker comes to his aid: instead of being hung, Jorgen is to be transported to the Antipodes for life.

On April 26, 1826, Jorgen arrives with the convict ship *Woodman* in Van Diemen's Land. The irony of his situation does not escape him. He has returned in a state of degradation to the very colony he helped establish twenty-four years earlier and is to be billeted in a town, Hobart, whose very foundation he helped to lay. Even worse: Hobart has not yet developed to the point where it has any gambling salons . . .

Jorgen's spirit is crushed, his wanderings pinioned to one

spare cushion on the global map. Home is the sailor, home from the sea, wrote Robert Louis Stevenson, and these words describe the later career of the former sailor and revolutionary, who, unfortunately, has found a home. He's found a job, too — Governor Arthur makes him a policeman. A policeman! The man who excluded the police from his Utopia! In his new role he hunts down bushrangers. He commands a unit of Governor Arthur's infamous Black Line and hunts down the last of the Tasmanian aborigines. He drinks.

Eventually Jorgen marries an Irishwoman named Norah Corbett and settles into a *ménage à trois* with her and John Barleycorn. All too frequently Hobart Town is treated to the spectacle of the ex-Protector of Iceland being chased through its streets by his drunken spouse. Or that spouse being chased by the drunken ex-Protector. To his credit, Jorgen doesn't auction off Norah to the highest bidder, which would have been considered quite proper in the colony. Instead he haunts the taprooms where his associates bow and call him "Your Icelandish Majesty."

Toward the end of his life he is greeted by a ghost, or the son of a ghost, from the past. Joseph Dalton Hooker, son of the botanist, shows up in Hobart Town with the HMS *Erebus*. In a few years the *Erebus* will be sailing into oblivion in the Canadian Arctic with her captain, Sir John Franklin, but now she's rigged out for a voyage of discovery in Antarctic seas. The young Hooker seeks out Jorgen and finds him at the taps, ill, destitute, and disheveled. When the young man identifies himself, the ragged figure seated at the table bursts into tears, after which he starts to ramble, on and on, barely coherent. Senile dementia, thinks Hooker, and writes his father that Jorgen is "quite incorrigible . . . irreclaimable . . . hopeless." Later the hopeless old man joins the cheering crowd at the docks as the *Erebus* sails off toward the last unexplored continent. How he must envy young Hooker his adventures! Two months later, on January 29, 1841, he raises his own anchor and falls dead in a roadside ditch. Perhaps his last thoughts have returned him to Iceland, the dog days of summer, and the good times.

Like Mozart, Jorgen Jorgensen lies buried in an unmarked pauper's grave.

Chapter 7

VIDEO NIGHT ON THE ARCTIC CIRCLE

UP UNTIL 1931 you couldn't get to Grimsey except by mail boat and then you'd have to wait six months before getting off the island, as this mail boat sailed only twice a year. *At most* twice a year. There is the story of a Danish tourist who landed on the island in the last century and literally could not leave it, for one storm after another roared down from the Arctic, cutting off all connection with the mainland. Finally the Dane had to give up. He married a local girl and (it's said) lived happily ever after.

Nowadays this northernmost fastness in Iceland is a little less impervious to visitors and, weather permitting, easier to leave once the visitor craves civilization's amenities again. Grim's Island (named for an eponymous tenth-century Viking rather than its ostensible grimness) is a half-hour flight from Akureyri by Twin Otter and only four hours from Akureyri by the ferry MS *Drangur*. As my original intention was to take boats whenever possible (otters, especially twins, ought not to have their frolicking impeded by humans), I took the *Drangur*, sailing with it up the long tranquil arm of Eyjafjördur, then barging and rolling over forty miles of blue polar seas.

The *Drangur* was not crowded. One islander was bringing

home his 102-year-old grandmother from the Akureyri hospital. The old woman seemed to be in great shape, reminding me of those sturdy New England *grandes dames* who are thought never to die but in the end simply dry up and blow away like autumn leaves. The other passengers included a day-tripping Danish birdwatcher, a honeymooning couple from Reykjavik, and a Grimsey man named Willard. Willard may not sound like an Icelandic name — it sits oddly on the tongue after so many Sigurdurs, Gunnars, and Thorgeirs. It isn't an Icelandic name. But it has been a rather common name on Grimsey ever since islanders took to naming their offspring after an American gentleman of means named Daniel Willard Fiske (1831–1904).

Willard Fiske was a traveler, a sports enthusiast, linguist, diplomat, journalist, and chess master. He wrote about Dante and Cornell University's Psi Upsilon fraternity with equal facility. His essays often took the road less traveled — one, for example, is titled "Syracuse as a Watering Hole." Another manages to find precursors of baseball in the writings of Pliny and Homer, and probably contains the first reference to the game as "the national sport of the United States." One year he is lecturing against the spoils system in American public office; the next year he has fixed his attention on the unhappy *fellaheen* of Egypt. His crusades on behalf of a new Egyptian alphabet ("One alphabet! One language! One country!") won him the accolade of an old *dahabieh* owner on the Nile, who said: "After Mohammed — Mr. Fiske!" In his leisure he composed chess problems.

As a student, Fiske had read the Sagas in their original Icelandic and always thought of the country as an island of rugged individualists rather like himself. In 1879 he decided to sponsor the installation of a telegraph cable that would plug Iceland in to the rest of the world. With a telegraph Icelanders would at least know when and if their sovereign, the king of Denmark, died (when Frederick VII had died during the winter of 1863, no one in Iceland knew about it until the first ship arrived on April 4, 1864). So Fiske sailed to Iceland to investigate the prospects for laying this cable. Along the way his boat passed a rough-hewn, uninviting little island at latitude 66°30′N, directly on the Arctic Circle. That's Grimsey, a deckhand informed him, and went on to say it was the poorest, most miserable piece of land in the country. Grimsey folk had the reputation of giving off an evil odor, the result of an exclusive diet of seabirds and their eggs. Winters were so cold that islanders refrained from sneezing for fear their noses would break off and go rattling across the floor. If pneumonia didn't take them, influenza would. If influenza didn't take them, the sea would; once all the male islanders drowned in a storm and the local minister was obliged to replenish the population through his own carnal enterprise.

Fiske already knew a little about Grimsey from his voluminous reading on chess. As Iceland itself had a passion for the game, even greater was the passion of little Grimsey. According to legend, the three-mile-long island had been settled by a clique of Viking wayfarers who spent their days and nights playing chess. One of them had the unfortunate ability to remember every move he'd ever made, until these moves clogged up his mind to the exclusion of all else and he slipped off into madness. Others took to their beds for weeks at a time in order to perfect a certain line of attack or endgame combination. It was not unusual for a man who lost a game through some ridiculous blunder to fling himself off the cliffs in despair. On occasion, so went the rumor, Grimsey folk still killed themselves over chess.

Poverty and chess! What better alliance for a man of Willard Fiske's kidney? The telegraph didn't work out, so he turned his attention to Grimsey. He was a quite beneficent man, about whom his good friend Mark Twain once said, "He was as dear and sweet a soul as I have ever known." When he got back to the States, he sent marble chessboards and chessmen to each of the eleven farms on the island. Later he provided Grimsey with firewood after the Gulf Stream no longer saw fit to supply its shores with mahogany from Honduras and palm trees from Haiti. In 1901 he underwrote all the costs for the island's first library. Upon his death his will divided his considerable estate among the Psi Upsilon fraternity, the Cornell University library, and this tiny outpost of the North.

Daniel Willard Fiske never once set foot on Grimsey.

Landfall. Grimsey was a rolling, knobby slab of land, folded and then folded again with harsh masculine grace. Treeless, windswept, rock-ribbed, hummocky, and bird-infested, it reminded me of Foula as well as certain islands I'd visited years ago in the west of Ireland. All its buildings — the shop-post office, the school-library-social hall, several fish-processing plants, and the usual ferroconcrete houses — huddled along its southern shore like survivors in the stern of a lifeboat. They were not old, yet they gave the impression of having had run-ins with rather a lot of bad weather.

An easterly gale was sweeping over the island's basaltic ramparts. I took refuge inside the shop-post office, where I pondered an impressive collection of videos, almost all of them American-made gore spectacles like *Thrúmar og Eldingar (Creepshow)*, *Stórislagur (Battle Creek Brawl)*, and *Mordhelgi (Death Weekend)*, though I did notice the alien presence of Olivier's *Hamlet* next to *Gereydandinn (The Exterminator)*. Well, not really that alien. There was the ghost and all those dead bodies at the end, not to mention poor Yorick.

Once the wind died down, I went back outside in search of a suitable campsite. Along the way I saw a little boy playing with a toy dump truck in the middle of the road. "*Hva heitur du?*" he asked me solemnly. What are you called? This has been the most popular question in the country going back to Viking times, when you had to learn if a stranger was kin and thus worthy of your stewardship or if he was an enemy and thus worthy of slaughter. Nowadays the question persists even though the consequences of answering it have, fortunately, died out.

When I told the boy what I was called, he gazed up at me with a look of astonishment, as if it were the first non-Icelandic name — apart from Willard — he'd ever heard.

I pitched my tent in the protective shadow of a mesalike knoll just north of the lighthouse at Ytri-Grenivik. An hour later I ventured forth to explore this remote little island of Viking chess players. A Baedeker I'd read in the Akureyri bookstore had dismissed Grimsey as "of no historical importance." When I read those words, my spirits leaped. For "historical importance," in the context of tourism, means ruins or palatial rubble, maybe a wheezing castle or a cathedral that's seen happier, more god-fearing days. I knew Grimsey would have the genuine article — *real* heaps of rock — rather than some man-made attempt at a facsimile.

The little boy was still seated in the middle of the road, still playing with his dump truck. Since we'd had our chat, not a single vehicle, apparently, had come along to dispute his right to play there. In fact, the only vehicles I'd seen on the island other than his toy truck were a few tractors and an antediluvian pickup, its fenders filigreed by salt.

"Hva heitur du?" he asked. I again offered him my name; this time he nodded sagely, as if he'd known it all along. Then his older brother emerged from their gray bungalow, and before *he* could get a fix on my name, I proposed a game of chess.

"*Skák?* They learn it in Akureyri, but we do not learn it in school here. Here we learn about volcanoes."

"But you don't have any volcanoes on Grimsey."

"Not now, but someday, perhaps . . ."

Almost immediately after I left the road, I heard the tell-tale cries and whoosh of wings past my ears that indicated I was being assaulted by arctic terns. They had a more expeditious method of attack than Foula's skuas, fierce but overweight birds that need a runway in order to take off. The arctic tern's lighter-than-air fuselage allows it not only to take off like a helicopter and execute STOL landings, but also to bomb you with winnowing acrobatics. It swoops down in almost no time, furiously slapping its wings and making loud irate shrieks. Hence its Icelandic name, *kria,* imitating its shriek, which it repeats over and over again — *kria! kria! kria!* — like a record stuck perpetually on a soprano's high C.

The birds had every right to behave aggressively: I was walking through their nesting grounds. But the arctic tern considers *everywhere* its nesting ground, so long as there is minimal vegetation (more vegetation would render its short legs inoperative). A bit of grass in downtown Reykjavik will do just as well as a field of stunted heather on Grimsey. The tern seems to dare the passerby to approach its little speckled eggs — eggs which, I might add, are to me the tastiest in the entire avian kingdom — and then attacks in a fashion that suggests it is tuning up its wings for the marathon journey south. Half of Iceland's arctic terns winter off the southwest coast of Africa, the other half in Antarctica. Those that winter in Antarctica travel farther than any other living creature during the course of a year. Such knowledge hardly consoles the person under attack: the arctic tern's little black-tipped bill occasionally draws blood.

I walked on, past tiny haycocks with sackcloth caps, into the increasingly bright afternoon. It was a curious brightness, at once sharp and extravagant but also a little off center, for in these northern latitudes the sun hangs low in the sky, and its rays always seem to have a hint of dusk about them. Maybe even a darksome hint of winter, too.

I turned north and passed rack after rack of fish hung up to dry — small mummies oscillating in the wind. Ten minutes later I walked onto a 3,000-foot strip of asphalt, which used to be a combination airstrip-chessboard, possibly the only one of its kind in the world. For a few days each summer, flights would be canceled and the best players in the country would convene here to push around life-sized chessmen. The practice died out in the early 1970s, just before the Fischer-Spassky match. A few dedicated Icelandic chess buffs wanted that match to take place on this airstrip, a location more in keeping with tradition than a big white modern dome in the suburbs of Reykjavik. Perhaps, but I wonder how Bobby Fischer would have reacted to his moves being relentlessly dive-bombed by squadrons of arctic terns.

At the end of the runway I noticed a road sign with arrows pointing in the direction of various cities (London: 1,972 km; Paris: 2,335 km; Rome: 3,436 km; New York: 4,448 km; Sydney: 16,137 km). The sign indicated I was standing directly on the Arctic Circle. Due north — over the Greenland Sea, the Lena Trough, and the Nansen Cordillera — lay the icepack of the North Pole, which was considerably closer than any of the aforementioned cities.

Beyond the airstrip I traversed grassy hummocks shaped into a carpet of camel humps by constant freezing and thawing. Soon this became a little tiresome and I bushwhacked over the height of land, Skollalag, to the eastern part of Grimsey. Dozens of popeyed sheep fled at my approach. Their coarse outer layer of wool looked like shingles on a roof and was by far the thickest wool I'd ever seen; a visitor to the Pole would not suffer much cold if he were dressed in one of these sheep.

Soon I was following cliffs beaded with the white scatter patterns of birdshit. I came to a big, round, precipitous bight where a man was poised at the edge of the cliff. At first I thought he was going to jump (a blunder at chess?), but when I got closer I saw that he was holding a long pole with a bowl attached to the end. He would ease this pole down the cliff face and bring it back up laden with birds' eggs. Most of these eggs he deposited in a basket; a few, however, he ate on the spot, throwing back his head as if he were downing shots of tequila. Before long he saw me. At my approach he cracked open an egg and offered it to me (*"Gerid svo vel!"*) much as another man would stand me to a drink. "I am a Socialist," he declared, adding, *"Skál!"*

I stared at the egg. Inside were the makings of a tiny kittiwake, two large eyes and wispy claws and veins sticking out from a reddish ooze. I stared back at the man, who informed me that the eggs with little birds in them were the best of all. So I said *"Skál!"* threw back my head, and thrust the contents of the egg into my mouth.

"Taste good?" The man smiled.

"Fine, but I could do without the claws."

In true Icelandic fashion, Sveinbjörn (for that was the man's name) began telling me about himself through his genealogy. His grandmother's people came from Snaefellsnes, just under the glacier into which Jules Verne sent his explorers en route to the center of the earth. He was a first cousin on his father's side to Einar Benediktsson, the poet who tried to sell the aurora borealis to an American businessman. Sveinbjörn's father was a fisherman and his mother a well-known healer. By the way, did I know that the human brain, dissolved in strong wine, would cure epilepsy? I didn't, I said.

Now Sveinbjörn told me he'd been a member of the Sósialistaflokkur since shortly after the Keflavík Treaty of 1946. This last bit of information, I knew, was being mentioned for my benefit. For socialists in already socialized Iceland have only one abiding issue — the NATO base. I braced myself for the inevitable.

"It is necessary for all the Americans to leave this country," Sveinbjörn said, lowering his pole and bringing up two more eggs. "I don't mean you, of course . . ."

"I know."

"We do not need 5,000 American soldier boys defending us against the Russian bear. How would you like 5,000,000 Icelanders in Washington, D.C.?"

"Actually, it might be rather fun . . ."

"*Já, já,* but your countrymen are no fun for us. When we have a nuclear war, this little country will be the first to be bombed, because of the American military here."

"So you think there will be a nuclear war?"

"Of course. It is prophesied in the old Icelandic rhyme, *Voluspá.* The god Surtur sets fire to the earth and heavens and everybody gets burned up:

> 'Black goes the sun the earth sinks in the sea
> no stars in the sky empty heavens
> only smoke in the sky and terrible fires . . .' "

Once the world had come to its ineluctable end, Sveinbjörn offered me another kittiwake egg. I shook my head and told him I was on a diet that expressly forbade me to eat more than one raw bird a day. Then he said:

"I will share my eggs with you, I will share my hearth with you, but one thing I will not share with you and that is my own little country . . ."

This sort of robust national spirit I'd encountered many times before. Iceland feels like a fortress and thus seems to excite a fierce guardianship in its people, who stand on the battlements with crossbows poised. Yet tiny Grimsey? There is so little on the island to spur the kind of passion that stormed the Bastille or dumped tea into Boston Harbor. Only cliffs draped with bird droppings, tedious hummocks, depressions, and stacks of columnar basalt. But love moves in mysterious ways. A barren place will never turn verdant on you and never change its face to suit some developer's grasping thumbs. The more barren it is, the more faithful its heart. Some of the saddest songs I know bid farewell to a faithful rock pile, but I doubt you'd get much sadness — regret, perhaps, but not sadness — in a song about exchanging London for New York or vice versa. Teeming crowds tend to compromise the affections rather more than nearly empty islands.

Now Sveinbjörn renewed his offer to share his hearth, along with a certain hearthside bottle of *brennivín* he kept especially for visitors. I'd also get to meet his daughter Ingibjörg and his little grandson Willard. Another Willard! At the mention of that name, I asked Sveinbjörn if he played chess.

"*Nej*. I only play poker."

July 4. An ink cloud of squall. A hemline of rain. Sleet. Hail. An ice-pellet shower. The northwest wind beats all and sundry against my shuddering tent with a muscular tattoo. I blundered out this morning to buy spirits for my stove and blundered back forty-five minutes later with my so-called waterproof polyethylene anorak a waterlogged mess and my flesh like jellied pus. The ground is a species of preternatural mud, with wind-rippled bogpools. I watch this wind blend a stream of my piss with the elements until my liquids and the Arctic's liquids are inseparable companions. Tentbound.

A tent's a miraculous invention by which you can heat the four walls of your shelter using your own quite inexpensive body. For the last two hours I've been sitting in here and jotting down random islets of thought in my notebook; the temperature and humidity are like Barbados. Try heating your own shelter — be it tar paper shack, condominium, or chateau on the Loire — with only the prodigality of yourself. Indeed, try folding up that tar paper shack, condominium, or chateau and securing it to your back. The building materials simply won't bend that way. (What's a house, anyway? Only a place where you don't have to eat your food on a stick . . .)

Despite the weather I'm a very curious item among the local kids. They slosh through the muck to visit my tent and then they ply me with the usual questions: "What are you called?" "Where do you come from?" "Is America at war now?" "Do you have a wife?" "Why aren't you working?" "Do you fish?" "Do you catch many fish?" "What *are* you called?" One gangly freckle-faced boy of ten or eleven solemnly squats next to me and pulls from his jacket a bottle of — cod-liver oil! He takes a few quick nips from it, puts it back, and then proceeds to tell me he can't wait to grow up so he can be a trawlerman, like his father. Norwegian-made stern trawlers are the best, he says, while Spanish trawlers are inclined to breakdowns and Polish B425s are completely hopeless.

I've dug up the following bit of advice from an old book of Grimsey folklore: If you want to be able to fly, you must obtain a bridle cut from the back of a recently buried corpse; you have to cut that corpse's scalp as a headpiece for the bridle, two bones from the head for the bit, and the hip bones for the bridle's cheek. Then you must find the seventh son of a seventh son and stick this improvised bridle in his mouth. Now you're ready to fly. Unless, of course, you've neglected to utter a series of magical oaths into the seventh son's ear, in which case all your efforts will be for naught and you won't be able to get off the ground.

No wonder those Twin Otters are so popular on the island . . .

By the next evening the rain had trailed off to a delicate mizzle and I went down to the pier, where I found Sveinbjörn and another man with a ten-foot Greenland shark. They'd just eviscerated it and flung its guts into the sea when the shark gave a powerful lurch and slithered off the pier into the sea.

For a moment it floated there benignly. But then it began to snap its jaws and devour its own viscera, a remarkable performance, I thought, for a creature that by all rights should have been dead. Sveinbjörn had to gaff the shark and hoist it back up again with a pulley. While he was whacking it over the head, he turned to me:

"Why not come and visit us tonight? I will share my house with you, I will share my food and drink with you, but one thing I will not share and that is —"

"Your own little country."

He nodded. Now they were removing big pale red chunks of meat from the shark's flank. They would take this meat, *hákarl*, and bury it in the ground for a few months to moderate at least some of its ammonia reek (Greenland sharks urinate through their skin pores), after which they'd hang it to dry in an airy shed. The longer it's hung, the more piquant the flavor. I know an old farmer named Eliás who insists that *hákarl* must be hung for a minimum of ten years or it's no good at all. Yet however long the meat is hung, it's still an unusually sharp and odoriferous item for the dining room table. Once when I was hitchhiking outside Reykjavik with a large piece of *hákarl* in my rucksack, the Englishman who picked me up kept casting dubious glances in my direction. He finally asked me whether I'd just robbed a grave.

Sveinbjörn said he'd join me after they finished with the shark, so I wandered up to his house and was greeted at the door by his daughter Ingibjörg, a lovely, honey-cheeked, flax-haired young woman whom I immediately wanted to abduct for my own nefarious purposes. *Já, já,* she said, you're the American who doesn't like little birds. Then she introduced me to Willard, her three-year-old son, and a friend from Akureyri named Heidbjört (Bright Sunny Day), who had two little daughters in tow. Heidbjört said she had once been to America, but the air smelled bad and she'd had to return home.

Here was an Icelandic commonplace — unwed mothers with children obviously born on the wrong side of the blanket. Elsewhere that might suggest the breakdown of a cultural system, but here it merely shows a wholesome interest in sex. As three quarters of all first births in the country occur outside wedlock, to call an Icelander a bastard would be a little like calling him a fish processor. A truism. I once heard about a girl in Húsavik who waited until she was married to have a child and her parents ostracized her.

Now Ingibjörg brought out a silver platter piled high with dried *steinbitur* ("stone biter," or ocean catfish), a tastier fish, to my mind, than cod or haddock and formerly more useful: distances used to be reckoned by the number of pairs of *steinbitur*-skin shoes a trekker over Icelandic wastes wore out. All of us dug in, including the children. I was especially taken by the sight of the two young women eating dried fish: they'd stick a piece of fish in their mouths, bite, tear, drag, pull, and rip off a shred of it, then crunch, crunch, crunch on it with their beautiful teeth.

"Would you like to see a movie?" Ingibjörg asked me. Before I could answer, she began rummaging among her cassettes. At last she pulled out an undubbed video print of *Straw Dogs*, showed it to me ("A pretty good film," I murmured), and inserted it in her VCR. At one time I did actually like *Straw Dogs*, but the events of the next few hours made me wish the producers had nipped the whole blasted project in the bud.

None of these Icelandic viewers could speak a word of English, but they all gazed at the screen like monks contemplating a palimpsest. I helped them with the plot, explaining, "That is a bear trap," "She is not getting enough sex," "They are nasty brutes," "They will give her sex," "He believes a man's home is his castle," "That man is the local half-wit," and so on. Every time the bear trap was shown, Willard would practice his first phrase of English exuberantly: "*Bear trap!*" And every time the Dustin Hoffman character appeared, Heidbjört's two little girls would yell in unison, *"Kani!"* (a derogatory word for an American).

This wasn't the first time I'd mediated between an American film and northern viewers. Once, in Angmagssalik, Greenland, I found myself in the position of having to present Woody Allen's futuristic film *Sleeper* to an audience of traditional Inuit hunters. The film was halted every fifteen minutes or so and I'd stand up and summarize the action up to that point, telling the jokes as best I could in Greenlandic. The audience was quiet, even solemn, while watching the film; they remained quiet and even solemn when I told the jokes, as if I'd been talking about Treblinka or Babi Yar rather than a supposedly funny movie.

I had no such problems on Grimsey with *Straw Dogs*. Everybody loved it. They loved the violence, they loved the body count, they especially loved the bear trap. After the last of the nasty brutes had been disposed of by the *kani*'s gun-toting wife, Ingibjörg said this final scene reminded her of the scene in *Njal's Saga* when Njal defends his home against Flosi and his band of marauders. Heidbjört said the idea of a bloody, overdone revenge reminded her of the Sagas, too. The children didn't say anything about the Sagas, but little Willard intimated that he wouldn't mind seeing the film again, as though he'd missed some of its subtleties the first time around.

So it was that we watched *Straw Dogs* all over again. "Bear trap!" exlaimed Willard. *"Kani!"* exclaimed the little girls. *"Hamra skal járn medan heitt er!"* exclaimed Heidbjört. (Kill the lousy bugger!) As for myself, the only thing that got me through a second showing so close on the heels of a first was the bottle of schnapps Ingibjörg had thoughtfully placed in front of me.

By three A.M. they were in the middle of a third consecutive showing of the movie. I was tired and bored, I'd finished the schnapps, and I was a bit irritated that Sveinbjörn hadn't put in an appearance. Just as I was getting ready to leave, he walked briskly in the door, an apologetic grin on his face and shark offal dripping from his trousers. He said he was sorry for being such a shabby host, but after they'd cut up the shark a herring boat had come in and he'd been asked to help unload it. Even now he was home only for a quick cup of coffee and then he had to go back and help unload crates from a second boat that had come in a few minutes ago. Would I still be around in, say, two hours? In two hours, I told him, I'd be in the depths of my REM sleep.

Back outside I was again dive-bombed by arctic terns, who seemed to be screaming, not *kria! kria! kria!*, but *kani! kani! kani!* Nevertheless I held my ground and watched the sun set and rise at the same time, a glowing orange fruit poised at the very edge of the horizon.

Toward the end of my stay on Grimsey I paid a visit to the local library, another staunch ferroconcrete structure modeled on a bunker or crematorium. But it did have one redeeming feature: a handsome mural on the outside wall that showed a fowler, his net on his shoulder, pausing for a moment of rest on the blue jaggedy cliffs of an island not unlike Grimsey.

I browsed through the library's holdings, which were still — even the Stephen King novels — being purchased from Willard Fiske's bequest. One entire shelf was taken up by books on Arctic exploration; another shelf housed the ubiquitous Sagas, which exist in so many editions (this library had three or four) that Icelanders tend to buy a particular edition to match the colors in their bedrooms or parlor, and the scarcely less ubiquitous *Likamsraekt med Jane Fonda*, of which there were three well-worn copies. Off in a corner was a pile of old chess books, accounts of epochal battles between long-forgotten masters and prolegomenas on obscure gambits, all gathering dust.

"Nowadays they don't play much chess on Grimsey," said the librarian, fetching snuff into first one nostril, then the other, elegantly, like a cat. He said one of Willard Fiske's marble chess sets used to be in the library, but it had been removed as an objet d'art to the National Museum in Reykjavik. He also said that Grimsey folk weren't reading as many books as they read in the old days, and he blamed this on the video craze. Right now, he thought, it might be a good idea to use Willard Fiske's money for the purchase of Home Box Office specialties like *Karate-Meistarinn* and *Besta Litla Gledihúsid in Texas*.

Before I left for the mainland, I visited Sveinbjörn's eighty-five-year-old father, Magnús. He took one long look at me and decided I was rather beaten down by my travels, so he cooked me up some seal meat, saying that seal has ten times more iron, per ounce, than cow. Then he told me a story about seals being descended from the Pharaoh's soldiers who had drowned in the Red Sea while pursuing the Israelites, and how certain people refused to eat seal meat because they thought they might be eating boiled Egyptian. Yet he couldn't tell me what had become of the chess set Willard Fiske a hundred years ago had sent across cold northern seas to his father.

Not being able to locate a single chess player on Grimsey, I assumed that the legacy of the Vikings who originally settled the island was dead. It was only upon my return to Akureyri that I managed to find a chess player, but he was a vacationing New Zealander and he beat me handily three games in a row.

Chapter 8

THE LIGHTHOUSE AT THE END OF THE WORLD

FROM AKUREYRI I hitched a series of lifts west to Isafjördur, a town of Day-Glo-painted houses huddled on the moraine of an old glacier. The afternoon I arrived, the temperature in Isafjördur, which is farther north than Nome, Alaska, was a sweltering sixty-eight degrees F and a couple of water-skiers were cutting wide swaths around the fjord as if it were Lake Como. Across this fjord, however, the peninsula of Hornstrandir was getting a fresh snow cover that seemed nothing short of miraculous, given the summery weather twenty-five miles away.

A big steepdown plateau of basalt, Hornstrandir sticks out from the cockatoo's crest of the Westfjords like a lone windblown hackle. Not only does it make its own weather, but it makes weather that is nastier than in other parts of the island. The peninsula is one of the last places in the country you'd pick for your saltwater farm, yet the handful of Viking colonists who settled here in the mid-tenth century had little or no choice in the matter, because nearly all the decent pasturage in Iceland was already accounted for. Also, a good many of these latter-day colonists were the Viking equivalent of Marielitos — cut-throats, ne'er-do-wells, and criminals — and they were not regarded as ideal neighbors by the rest of the population.

One of these colonists was Eric the Red's father, Thorvald Oskarson, a murderer outlawed from his native Norway. Thorvald came to southwestern Hornstrandir and planted his chair pillars on an interim of flat land he called Drangur (Lonely Rock), after a tall sea stack near his holding. Drangur offered him the chance to go straight if only because it was so isolated: On whom could he now test his murderous proclivities? On the other hand, Thorvald's red-haired son Eric detested Drangur and could hardly wait to leave its sun-denying mountains and encompassing drift ice. Finally Eric married his cousin Thjodhild and moved to her family's holding in Snaefellsnes, which, by comparison with Drangur, must have seemed like the Good Earth itself. Unfortunately Eric murdered a man, Sour Eyjolf, who had murdered some of Eric's slaves, and he was banished from the district. Later, on an island called Öxney, he murdered another man (he was truly his father's son) and was compelled to leave Iceland altogether. He set sail for Greenland, but that's a subsequent story.

Though Eric left, a number of his Viking countrymen stayed on in Hornstrandir for a thousand years, mingling their lives with the astringent salt of Denmark Strait. They fished, farmed, fowled, and gathered driftwood washed up on the peninsula's bellicose shores from the great rivers of Siberia. Only in the 1940s did the antique fabric of life begin to rift apart, mostly because the fishing was inshore, done in open boats beneath two-thousand-foot cliffs, and it could not compete with the big trawlers that were starting to dominate the Icelandic fishing industry. First one family departed, then another; some floated their houses intact to Isafjördur, others settled in Reykjavik. By 1959 the last family had packed its bags and Hornstrandir was once again pure and unsullied wilderness, as remote from the quotidian life of the human species as a pebble on Mars.

I had one more thing to do in Iceland: to visit Hornstrandir and call on the very last person — a man named Jóhann Pétursson — who can claim the peninsula as his place of residence. For twenty-six years Jóhann has been the lighthouse keeper at the North Cape outpost of Hornbjarg, ringing up Reykjavik every few hours with some new tidbit of unpleasant news about the weather. None of his assistants had lasted longer than a few months. I had heard that one woman showed up for the job with her entire wardrobe, expecting to trip the light fantastic at local dances. But the nearest local dance would have required her to trek seventy miles over rough mountain paths, for which she would need Vibram or Galibier-soled boots, not heels. So she simply stayed home (and went back home, the first chance she got, to Reykjavik).

The day after my lift deposited me in Isafjördur, I took the *Fagranes* ferry to Hornstrandir. The peninsula still has a seasonal boat service even though it doesn't have any people, a conundrum explained by the fact that many of its former residents or their offspring still maintain summer houses — modest dwellings that resemble corrugated iron privies — in former settlements like Hesteyri and Adalvik. No doubt they'd hop in their Moskvitches and Skodas and drive to these houses if they could, but there isn't a road, just a few thin arterial footpaths, so they sail over in this brown shoebox of a boat, its gunnels so low-slung you can reach down and snatch a jellyfish from the water.

July 11. Another sunny day. We chug along Snaefjallaströnd, Snowy Mountain Beach, past abrupt purplish mountains tossed up during the Eocene and then sheared flat by Pleistocene glaciers. Their clean terraces remind me of the multilayered petticoats of the last century, except for a few quite unladylike piles of scree. A dark-haired ("My people married their Irish slaves"), stout-waisted man named Sigurdur informs me that

Snaefjallaströnd used to be called Snaefjalla*draugur*, Snowy Mountain Ghosts, since it had such an abundant population of ghosts. He insists they really were ghosts, not figments of the imagination or curious sculptings of rock, and cites as evidence this story: When he was a boy, he was searching one night for a lost sheep and happened to meet a very tall man, gray of face, who looked hungry. He offered this man some of his *hangikjöt* (smoked lamb) as well as a knife to slice it. *Nej, thökk fyrir*, said the man, taking off his head and tilting it politely, We ghosts don't need knives. Then the tall gray man proceeded to eat the entire leg of *hangikjöt*, after which he pressed a seventeenth-century silver crown into Sigurdur's hand and wandered off into the night. I still have that crown, Sigurdur says.

In Jökulfjördur, Glacier Fjord, fleets of puffins float on the water like clergymen at their ease. One of them whirs past the *Fagranes* with so many sand eels tucked in its beak that it appears to have sprouted a gigantic drooping mustache, like Nietzsche's. We pause at a bight called Grunnavik, where Sigurdur and his wife slosh ashore in waders, followed by a relay of their provisions. Grunnavik has three small cabins (just the right size for a town) and mountains with such precise ridges they could have been laid out by geometers — or petticoat manufacturers.

Northeast into Hrafnfjördur. An old Westfjords man once told me that one day in the 1940s he was fishing here when he noticed a colony of quite unusual-looking birds perched on a grassy cliff ledge. Through his binoculars he saw that these birds were nearly three feet tall and had an erect stance, huge beaks, and very tiny wings. At first he took them for penguins, but he'd never heard of penguins in Iceland before, so he wondered whether they might not be some sort of mutant puffin, grown to an uncharacteristic size. Back in Isafjördur he called on an ornithologist who showed him a book of Icelandic birds.

The book didn't have a photograph of his bird, so the ornithologist brought out another book and the man browsed through it until he found a drawing of the *geirfugla,* or great auk. *Já, já,* the man exclaimed excitedly, that was the bird he'd seen. Impossible, said the ornithologist. The last pair of great auks — the last pair in existence, probably — had been clubbed to death in 1844 on Eldey rock just south of the Reykjanes peninsula. There hadn't been a single authenticated sighting since then. But the man kept pointing to the picture and exclaiming, "*Já, já,* that's the bird I saw." A few days later the two of them went to the cliff where he'd observed the colony, but the only birds their binoculars could make out were the usual puffins, murres, and razorbills . . .

. . . And now these are the only birds my own binoculars can make out, too. But maybe great auks don't like binoculars. Maybe they don't like people, either. If I were one of these big, flightless, fine-tasting birds — hunted down by Neanderthals, Vikings, Celts, and Grand Banks fishermen alike — I doubt I'd expose myself to human-type beings regardless of how civilized they might appear. Public appearances only land you in the stewpot.

Now we draw up to a spit of a moraine and Captain Örn Örnsson, Eagle, Son of Eagle, gestures toward a congeries of scarplike mountains to indicate the overland route to Hornbjarg. It's two days at most, he says, unless, he adds with typical Icelandic wit, I fall down and die, in which case it'll take somewhat longer. He instructs me to follow the cairns to Skorardalur, hike along Skorardalur to Furufjördur, then follow the skein of abandoned farms northwest to the lighthouse. If you meet a polar bear along the way, he tells me, back up slowly, singing the Danish national anthem at the top of your voice. *Já, já,* that gets them every time.

Alone at last! Long cirrus clouds put me in mind of what it must be like inside the skeleton of a dinosaur. I clatter east along a shingle beach darkened like moleskin and chiseled to grit by the quicksilver surf. Soon I see a dead seal, its intestines flowering in the chill sunlight, strangled by monofilament and partially eaten by (I'd guess) arctic foxes. Then I see a dead fox, eyes picked clean, flesh ripped asunder, one ear gone — the work of birds. Also a few birds themselves, their frail cages flexed at grotesque angles, as finespun as spider's floss. Twenty miles from the Arctic Circle it's an edible world — killer and victim, predator and prey, maggots and I, all dance together in the never-ending *pas de deux* of hunger.

July 12. Dawn is the soft dark red of raw liver. I rouse myself to wakefulness with a trip to the local public house, a hullabaloo of a mountain-run stream whose mineral water has the heft of a full-course meal. In the book of Judges, Gideon picked an army to defeat the Midianites by leading his prospective candidates down to a river; those who stuck their heads in and slurped the water like wild animals he chose over those who gently cupped their hands and lifted the water to their lips. I would have been drafted, I fear, in the very first round.

Soon I'm walking through mossy, boggy ground that alternately sucks and sighs at my boots in a two-note beat that persists — SLURP! AACCH! SLURP! AACCH! — with each step. Then I cross a series of snowfields as white as freshly laundered linen, laid out along a crinkly ridge that probably hasn't seen the sun for years. On higher ground I enter a meadow of infinitely delicate star-shaped flowers, tufts of lady's bedstraw, and buttercups. It's not exactly the Amazon rain forest, but it's not a cinder-heap austerity, either. Just goes to show how pleased the slender soil is not to have all those tireless sheep nibbling everything down to the roots. There's angelica everywhere, covered with buzzing flies, and big ruddy "berserker" mushrooms — a mildly hallucinogenic mushroom the Vikings ate to make themselves combat-happy. (I once sautéed up a few and they gave me double vision for an hour, that was all.)

Onward. Near a heap of licheny detritus I see an arctic fox, blue-gray with honey-buff underparts, clutching a bird's egg so carefully he might have been soft-mouthing a Fabergé. He's moving along at a kind of spunky, egocentric trot, as if proud of his ability to carry an egg a couple of miles and not break it. Come winter he'll be wearing this same somewhat sooty coat; foxes who raid bird cliffs stay "blue" the year round, whereas those who get their rations in the interior sustain rather worse weather (the Gulf Stream doesn't wrap around their dens) and usually turn almost pure white.

Once I reach Furufjördur on the north coast, I begin to notice more foxes, so many and so vocal, barking doggily to each other, that it's like walking through an enormous open-air kennel. The kits behave just like puppies. Up and down scoria slopes they trail me, edging as close as my shadow, yiping and snorting, trying to figure me out — a moving boulder? a leviathan puffin with a backpack? It's been thirty-five years, or several vulpine generations, since their antecedents in Hornstrandir were marked animals, shot for their pelts or for their habit of going after sheep with the ferocity of wolves, flying at the throat and hanging on till the sheep dropped from exhaustion. Thus these young foxes probably don't know the truth about me, namely, that my kind has been trying to decimate their kind ever since my Viking predecessors came to these shores (the Vikings were the second mammal to colonize Iceland; the arctic fox was the first). Just north of Furufjördur I'm trailed by a trio of young foxes who must think I'm something they can practice their hunting on, a sort of scarecrow prey. They keep trying to herd me into cul-de-sacs, which I let them do, but when I lie down on the ground and play dead, they gaze at me sullenly, as if to say What the hell, man, how come you won't play our game?

Nobody who has had any dealings with the arctic fox doubts

its canniness. An old hunter named Thordur once told me he'd seen a fox back up against a roost of kittiwakes, its brush held aloft, so that the kittiwakes would think this newcomer was just another big white bird rather than a pillager of their nests. The last time I saw Thordur, he was getting ready to climb up into the Snaefellsnes mountains and he had a hum of perfume so strong it just about lifted me out of my chair. He was convinced the foxes knew his scent and even had a particular bark to notify each other it was he, Thordur, and not some other hunter coming after them. So he'd covered himself with Chanel No. 5 to fool them. In a couple of weeks, he said, he'd have to switch to Old Spice or maybe seal grease once they caught on to Chanel No. 5.

Up and down, up and down, through country that would make a first-rate roller-coaster course. Every time the land dips to the sea, I'm obliged to wade across a knee-deep river so cold it sends knifelike prickles of astonishment along my bare legs. Near these rivers there is always an irretrievable farm or, more accurately, a pile of rotten planking and a few rusted-out scraps of metal, perhaps a coil of chicken wire, all that remains of an age-old sufficiency. At one of these farms I rest on a wormy old bench carved from a Siberian birch log and eat a half-melted bar of Prince Polo. Make that two half-melted bars of Prince Polo. Necessary fodder in these relentlessly steepdown parts, for chocolate replenishes lost muscle glycogen (high-protein grub like nuts and cheese drains the muscles), digesting quickly to keep the level of blood sugar perking, which in turn keeps the brain perking, alert to its own insignificance in this largesse of abandonment.

It's been an unusually lucent day, sharper as well as more genial than I expected, and for an hour or so I managed to persuade myself that I could see the Blosseville Coast of East Greenland, a range of snow-cased mountains two hundred miles to the west. Finally I realized I was a victim of the *hillingar* effect, an Arctic mirage in which air masses are inverted and cast in the shapes of mountains. Except that these were such comely, well-proportioned mountains, I didn't really feel like a victim at all.

Evening. Halted six miles from my destination in a pocket-sized cove so perfect in its solitude I'd have felt bruised to fling a word or two at a passing stranger.

Toward noon the next day I climbed a steep scree-laden slope above the derelict farm of Bjarnarnes and descended to Hornbjarg along a route of low slanting hills made up of alternate layers of basalt and reddish tuff. At last I reached a bowl of a valley so bursting with angelica that the flies in all those flower clusters sounded like a plague of locusts. Half a mile north the cliffs bulged out like huge insulting chins, dropped into the sea, and then gingerly returned as a row of fissured stacks, accusatory fingers of rock pointing skyward as if to say: Don't blame us for your troubles, mate. Blame *Him*.

The clean white lighthouse and yellow beacon tower rested on a spit of basalt fifty or sixty feet above the roistering sea. Its keeper, Jóhann Pétursson, was reading the wind gauge when I arrived. He was a nimble, red-haired, bespectacled man, perhaps sixty-four or sixty-five, whose strong-featured face suggested that he would read that blasted wind gauge even if he had to climb twenty miles of acclivitous mountains three times a day to do so.

Jóhann talked freely and easily about himself, having apparently lost the distinction (like a lot of other solitary people) between hard facts and intimate details. He told me how he'd grown up in a very strict, old-fashioned family in Grindavik, fifty miles south of Reykjavik, and how he hadn't slept with a woman until he was twenty-seven, a nearly geriatric age for an Icelander. As a young man he'd had so many fears and bugaboos — women, ghosts, bad weather, his own dreams —

that he was terrified of being alone. Being alone caused him to break into a sudden, cold, drenching sweat. It was not unusual for him to sleep with a gun under his pillow and wake up with it clenched tightly in his hand. Then one day he was walking along the shore at Grindavik when he happened to stumble, literally, on a dead body. The body was a lascar's and it had come from a 20,000-ton tanker that had broken up on the rocks off Grindavik a few days earlier. He rang up the company in Reykjavik responsible for the tanker, but they didn't seem interested in the lascar, so he had to make a coffin for the man himself. This he did, though he was so exhausted from his efforts at carpentry that he fell asleep with the coffin beside his bed. Next morning he awoke with a strange elation rather than his usual sweaty fear, as though he'd finally seen the skull beneath the skin and discovered that it wasn't so bad after all.

Jóhann took the Hornbjarg job in 1961 because he figured it would buy time for, and even fuel, the novel he was writing. It was unheard of, he said, for an Icelandic writer to combine teaching with the labor of his pen. To do such a thing ("I hear all American writers are professors. How dreadful!") would be like putting a leaden weight around his inspiration, he thought. His literary colleagues tended lighthouses, from which they still managed to carry on a lively dialogue with their public, like the Skálavik keeper Oscar Adalsteinn Gudjónsson, who read sections of his works in progress over shortwave to the fishing fleet. Yet between navigating boats around icebergs, gathering errant fishing floats, and enduring assistants who couldn't read the cloud charts (like, for example, his present assistant), he, Jóhann, had scarcely written a single word in twenty-six years. *Já, já,* he grinned, there seemed to be a leaden weight around his inspiration anyway.

He had come to the lighthouse fully aware of the so-called curse, which he believed in and didn't believe in, both. It seems the last farmer at Hornbjarg had stood on the scarp where

Jóhann himself stands to observe drift ice and had hurled down all kinds of imprecations on his land. This he had done a few years before the lighthouse was built in 1934; since then all fourteen of the keepers before Jóhann had sustained some kind of mishap, from injury to unexpected illness, which kept them from staying on past their fifth year. Toward the end of his own fifth year Jóhann had been hunting birds on the cliffs when he felt someone or something give him a push. He toppled down twenty feet onto a ledge. Somehow he managed to drag himself back to the lighthouse and ring up the coast guard, which airlifted him by helicopter to Isafjördur. There he lay in the hospital, suffering a medley of bruises, loss of equilibrium, and a very bad concussion, wondering if the fears of his youth would resurrect themselves and intrude upon his destiny with Hornbjarg. They didn't.

"Don't you feel lonely here?" I asked him, noting that he was the only year-round resident for seventy miles with fewer than four legs or less than two wings.

"Look at those lovely cliffs, my friend! Look at that mist on the ground! Listen to the rush of that stream! How can a man be lonely with such companions?"

We went inside the lighthouse and I saw another, equally good reason for not being lonely. Here in the Westfjords it used to be said that the more remote the holding, the bigger the library and the more literate the occupant. Well, Hornbjarg was about as remote a holding as you could get without slipping off into irrevocably Arctic seas, and by Jóhann's own estimate the lighthouse had 16,000 books lining its walls, rising up in precarious piles from the floor, thrust into boxes in closets, and choking the cloud charts. (Was that why Jóhann's assistants couldn't read them?) I even expected to find a few books inside the freezer next to his cache of whale meat.

Pride of place went to the Sagas, complete in several editions, including a few odd volumes of *The Foster-Brother Saga*, which is set around Hornbjarg and is best known for a scene where one fowler falls off a cliff and another fowler tells him not to make so much noise, he'll scare the birds. But unlike some of his countrymen, Jóhann did not draw the line with the Sagas (or the Sagas' apparently universal shelf-mate, Jane Fonda). I also noticed the collected works of Faulkner, Dreiser, Hemingway, Dostoyevski, Hamsun, Hardy, Nexø, Dickens, Isak Dinesen, and the Icelandic Nobel laureate Halldor Laxness. There were well-thumbed copies of Dee Brown's *Heygdu mitt Hjarta vid Undad Hné* and ex-Chairman Mao's *Rauda Kverid.* There was the Icelandic translation of *Moby Dick* done by Julius Hafsteinn, a retired sheriff from Húsavik, who had seen two whales copulating at sea and was so impressed by the sight he decided to translate Melville. There was Joy Adamson's *Borin Frjáls*, John Steinbeck's *Thrúgur Reidarinnar*, Richard Llewellyn's *Graenn Varstu, Dalur*, the midwife Margarete Tómasdottir's translations of Colette, and the pharmacist Helgi Hafdanarsson's translations of Shakespeare. There was an odd quarto volume in English entitled *A Short Commentary on the Flowing Back of the Waters of the Red Sea for the Passage of the Israelites.*

Jóhann had the most eclectic reading taste of any person I'd ever met. I tried to convice him that *Sámsbaer (Peyton Place)* wasn't a very good book. ("You actually like that book?" "I do." "But it's very badly written . . ." "Not in Icelandic, it isn't.") But I didn't persist, since a hungry man ought to be allowed to consume whatever he pleases.

Said Jóhann: "Did you ever read *Ástsaga* [*Love Story*]? Now that's another good book . . ."

And maybe it was, in the lovely mists of Hornbjarg.

For the next few days I remained at the lighthouse, walking in the mist, catching up on my reading, and listening in on fishermen's ship-to-shore calls courtesy of Jóhann's radio-telephone. ("Are you by yourself? Are you *sure* you're by yourself? What's that voice I hear in the background? It sounds like a man . . .") One day I encountered an arctic fox so unaware of human belligerence that I was able to feed it half a Prince Polo bar out of my hand (like a true Icelander, it seemed to love Prince Polo). Another day I wandered down by the shore and beachcombed a truly prodigal collection of flotsam and jetsam, carried here by northeasterly winds. That same evening I told Jóhann about a drift can of Heinz tomato juice I'd found with six brittle starfish inside it, twined together in a knot. He figured they'd chosen that particular can because they were red and it was red and thus it must have seemed like good camouflage. Then he showed me a Dutch wooden shoe he'd picked up a few years back at the same spot where I'd found the tomato juice can. It was discolored and badly worn, a mere shadow of its former self, having traveled by God-knows-what route from Holland to Hornbjarg.

"I plan to stay here until the other shoe washes up," Jóhann told me.

"What if it takes the rest of your life to wash ashore?"

"*Nej*, I can't wait that long. I must retire from the lighthouse at the age of seventy."

And suddenly, at the prospect of exchanging the lighthouse for the metronomic bustle of Reykjavik, his face clouded over and he looked like the loneliest man in the world.

The following morning the coast guard supply ship paid its monthly visit to Hornbjarg, and I talked the captain into letting me sail with him back to Isafjördur. Strapped into the Zodiac, which leaped against one wave after another en route to the ship, I remember thinking that the five of us — the captain, three crew members, and myself — were rather a lot of people to be congregating together at one time, in one place.

Chapter 9

A WALK ON THE WILD SIDE

IN ERIC THE RED'S TIME, you could hop an open-decked merchant ship or possibly hitch a lift with a Norwegian bishop for the five-day run to Greenland's Eastern Settlement. Nowadays the connection isn't quite so easy, as I learned when I wandered the Reykjavik docks in search of a Greenland-bound boat. Neither cargo ships nor passenger ships like the *Norröna* made the run between the two islands, and trawlers weren't supposed to take on travelers like myself. A skipper who was a friend of a friend offered to bring me along as a paying stowaway, but he warned me they might not actually land; last time out, they caught their cod and steamed directly back to Iceland, much to the disappointment of the crew, for whom fishing off Greenland often meant a visit ashore to sport with native girls.

Finally I realized I'd have to fly if I wanted to reach Greenland before the onslaught of winter. By flying, I would be violating my original plan. The Vikings probably wouldn't have flown even if they'd had the chance. They were a sea-going, sea-hearty people who always traveled with their worldly goods; doubtless they'd have balked at the prospect of herding their cattle into an aircraft. On the other hand, native Greenlanders have always used the jet stream. Stories about their inspired aviations abound. An old *angakok* (shaman) named Angerlerduaq once took to the air by rapidly beating a pair of guillemot wings over his head, and two other *angakut* named Ayiyaq and Uitsaleqangiteq used to stage aerial competitions for the benefit of spectators. And the mountain hermits known as *qivigtut* were able to fly simply by placing their thumbs in their nostrils and wiggling their index fingers; once aloft, they would float one leg behind them like a rudder and use the other to steer. Robert Petersen, of the Inuit Institute in Nuuq, once told me he'd heard about a man who could fly just by bending one leg slightly and flexing his little fingers back and forth — a major breakthrough, Petersen thought, in the science of aerodynamics.

And so it was that I went to Greenland more or less in native style.

On previous visits I had fallen crazily in love with this capacious island, shaped like a colossal stockinged foot, with a land area four times the size of France and a population slightly smaller than Wauwatosa, Wisconsin. I had fallen in love with its high-flying natives, a race of people who call themselves Inuit (Real People) or Kalladlit (Real, Real People), but never Eskimos, which is a Cree Indian term of disdain meaning "eaters of raw fish" (Frenchmen don't like being called "eaters of garlicky snails," either). I had even fallen in love — for a night — with a young woman who possessed by far the finest pair of eyes I'd ever seen, thus confirming an old northern seafarer's adage that the only two beings in all the world with human eyes are seals and Greenland girls.

I'm not usually inclined to betray an intimacy, but I feel obliged to say a few words about this last-named love because it was as much a cultural liaison as an amatory one:

I had gone to the tiny East Greenland village of Igateq to

meet an old man named Salluq, reputedly the last impenitent heathen (others had converted to Christianity by the late 1920s) in the country. Unfortunately Salluq was sailing from Igateq to Angmagssalik with a cargo of sealskins just as I was sailing from Angmagssalik to Igateq. We never met. And now I found myself stuck in Igateq. The village was draped along a rocky eminence whose hard-used beauty lay wild around it like contours on a map (the perfect place, I thought, for a last heathen). I pitched my tent in the lee of this eminence and settled in for the evening. I was just making a good gut-cauterizing pot of coffee when a young woman suddenly appeared beside me. According to the explorer Knud Rasmussen, Inuit women move with the silence of the ages, and that's exactly how this woman had moved — I hadn't even heard a crunch of silt and gravel to indicate her approach. She was just there, as if by magic. Now she smiled. I smiled. She smiled again, saying: "*Kusunsuaa unsukkapiq?*" (Do you have a nice penis?) I was a little taken aback, as this wasn't the sort of question I was accustomed to in my country. Here, the woman explained, "a nice penis" meant I didn't have gonorrhea. I said my penis was fine, yet I must admit I had no idea what her mission was. Perhaps she was the village nurse carrying out some sort of medical survey.

The woman departed and an hour or so later came back with her husband, a man whose neat tapered body seemed ideally designed to fit into a kayak. He grinned and whacked me on the back, spilling half my cup of coffee. *Ela!* he exclaimed, My little wife fancies you. Glad to know it, I mumbled, still a bit confused, wondering why she fancied a fleabag camper like myself. And, he added, I would be honored if she and you had sexual intercourse tonight. At which point I nearly spilled all the rest of my coffee, too. *Tonight?* I said, buying time to think it over. He nodded pleasantly and whacked me on the back again. I peered up at his little wife, then back at him, and — being young, adventurous, and perhaps even a little naive — agreed to take him up on his very generous offer.

And so I was introduced to the Inuit custom of wife swapping in its modern incarnation. In the old days the custom satisfied bodily desires even as it kept the wolf from the door. If a man loaned out his wife, he could expect to get meat from her lover once his own cache of meat was gone; the resulting tie was as deep as a kinship tie, and lovers who didn't share their meat often died under very mysterious circumstances. Today, however, there's no need to share one's comestibles — a person can always drop by the Royal Greenlandic Trading Post and purchase a few tins of South African pilchards or a plasticized smear of Danish ham. Yet Greenlanders remain free and easy with each other's charms, as if their mind-set had not quite caught up with the times: I was being *given* a wife with no strings attached. Whether it was because I appeared to be lonely or the woman simply liked my looks, I never found out.

We went back to their two-room shack and I spent the night with the woman Katrina's brown, wondrous eyes gazing up at me and her smooth flesh blended into mine. It was the sort of night when every part of your body has a sense of belonging to every other part as well as to the naked globes and grasses of the carnal Universe itself. But what I most remember about that night was the husband seated at a table in the next room, cheerfully cutting cards, CLACK! CLACK! CLACK! over and over again. Next morning he thanked me for my visit and gave me a send-off in the form of a succulent haunch of seal.

I returned to my tent and didn't see either of them again. Several days later I was in an outboard heading back to Angmagssalik when the man at the throttle turned to me and inquired whether Katrina made love softly, in the Greenlandic manner, or noisily, in the manner of Europeans. I hesitated and then said: In the Greenlandic manner.

But Greenland itself. I would be a rather shameful lover if I blinded myself to its enormous swarm of problems, not least of which is the apparent irreconcilability, inside and outside bedroom walls, of Greenlandic and European manners. For three hundred and fifty years the island was a perennially underachieving colony of the Danish crown. While African and South American colonies were working overtime on behalf of their European proprietors, Greenland managed only to be a drain on the Royal Danish Treasury. Yet the Danes persevered, if only because they thought there had to be something akin to a resource under all that ice; even then they were less greedy than most colonial proprietors. They dragged their huge woebegone charge from one age to the next, providing social and medical services, religious instruction, and the best European education a non-European could possibly desire. With the best of intentions, they made European victuals available to all, from canned foods to sugary glop, not realizing that imposing a shift in food habits would leave a subsistence-fueled people with not much to do anymore. And with slightly less than the best of intentions (they still hoped there'd be *something* under the island's interminable ice or in its unfathomed seas), they gave Greenland a sort of phantom Home Rule in 1979, retaining for themselves control of all preeminent policy decisions. Greenland is free; the Danish raj lives on.

Our plane made a very spirited landing at Narsarssuaq, the only airstrip in southern Greenland long enough to receive a fixed-wing aircraft. Many of my fellow passengers were Icelandair employees who showed neither the patience nor the steadfastness of their ancestors. Whereas the Viking colony had lasted five hundred years, they planned to stay in Greenland only half an hour. They had no interest in the island itself; their plan was to fly directly back to the Keflavík airport and clear out a few shelves of duty-free booze, available only to international passengers, which, in their transit from Iceland to Greenland and back again, they would become.

I could hardly blame the Icelanders for choosing to leave Narsarssuaq the instant they arrived. It was a disagreeable place — just a runway and a hotel at the head of a long claustrophobic valley. Once a U.S. air base, it became a demobilization hospital during the Korean conflict and quartered the saddest and most pathetic of the war wounded, men whose viscera had been blown away or whose faces had been left behind on the battlefield. Most of them died, and their bodies were cremated and ashes scattered in the glacial silt of the so-called Hospital Valley. A friend of mine once visited here and came across the brainpan of a man who doubtless did not expect to end up in Greenland when he signed on for Korea.

As I stepped into the warm sunny afternoon, I was greeted by the local welcoming committee, taking time off from their busy schedules to salute my arrival. Perhaps they had just been mobbing a musk ox to distraction, or forcing a reindeer to make a suicidal leap from a cliff top, or taking their time with the murder of a brooding bird or two. Whichever, they were now giving me the full-bodied benefit of their attention. I refer to that tiny celebrant of all warm-blooded life, the mosquito (*Aedes rearcticus*), whose remorseless presence makes much of the west coast of Greenland buggier than Amazonia.

On warm days like today, *Aedes* rise up from bogs and willow thickets in enormous, noisy clouds. The males are gentlemen, hovering lazily and buzzing each other with stray tidbits of insectual wisdom, like perpetual bachelors at the firehouse. But the women are whining peevish viragoes capable of penetrating jeans, underwear, and skin with a proboscis as sharp as a needle and as subtle as a drill bit. Bloody persistence! Once I even saw a female *Aedes* probing for a capillary through an eyelet on my hiking boot. Yet the victim deludes himself if he thinks his hemoglobin is the staff of their life. Not true at all. The arctic mosquito can get all the sustenance she requires from the juices of willow herbs, but she needs blood to condition her reproductive equipment and so allow her to become a mother to hundreds more like her.

In a 1966 article in *Meddelelser om Grønland,* Danish scientist Erik Nielsen described how he once counted a landing rate of eighty-three "individuals" per minute on a person — himself — sitting in the shade at Narsarssuaq. Just now, as I stood in the sunlight, this did not seem like very many. I was encompassed from head to foot and naturally tried to reduce their numbers by slapping a few of them to perdition. The more I killed, the more rose up in apparently endless maternal relays. As genocide didn't seem to be the answer, I decided to show a little respect by rubbing on some Tiger Balm. A good many Greenlanders swear that this Singapore-made panacea for rheumatism, headaches, and gout is the best mosquito repellent on the market — much better than seal oil or overdose quantities of vitamin B. Others think repellents only make a bad situation worse. Tiger Balm does stop mosquitoes from eating you alive, but it doesn't stop them from getting stuck in your ears and nostrils, stepping into your eyes, getting tangled in your hair, and jumping to their dooms down your throat. Hold the Tiger Balm and they'll just suck and depart. Apply the Tiger Balm and you'll get a swarming, gossipy mass of mosquitoes as everlasting consorts, along with the odor of camphor.

My present destination was Igaliko, former clerical headquarters of the Vikings (Igaliko means "Abandoned Cooking Place" — the Greenlanders who named it thought empty pots more eloquent than an empty cathedral) and point of embarkation for eleventh-century trips to Labrador and Newfoundland. The boat for Igaliko did not leave until the next morning. As I had no desire to linger around the airport or take a room in the overpriced barracks calling itself the Arctic Hotel, I walked toward Hospital Valley. What I saw along the way was not a pretty sight.

Narsarssuaq lost its usefulness in the mid-1950s, after the Army Corps of Engineers built the Thule Air Base to provide American paranoia about the Soviets with an outpost in the Far North. The GIs left in 1959, and the Danes sold much of the scrap from the base to a Norwegian scrap merchant — much, but not all. There is still enough heavy metal lying around to make Narsarssuaq a first-rate tourist hangout for William S. Burroughs: a terminally rusted B-52 resting on its back like a fallen pterodactyl; military vehicles stripped to their essential nakedness; the corpse of an old asphalt factory; vestiges of antiaircraft batteries; snakelike fixtures with coiled necks rising a foot or two off the ground; and convoluted bits of machinery whose only purpose would seem to be their own purposelessness.

I had no interest in camping among all this dead technology, so I continued walking for several miles along a pebbled glacial plain carpeted with gray-green moss. I walked until I came to a river so silted up it looked like chocolate milk. On the bank of this river I pitched my tent, with only a few "rogue" mosquitoes (it was now too cool for most of the little fiends) to gnaw at my skin and nerves. Directly ahead of me stood Qorqup Sermia, a tongue of the ice cap that resembled a long rough-fronted wall. Directly behind me was the jungle, thick copses of dwarf rowans and dwarf willows with seedpods like delicate gray earrings. A few bluebells lay scattered about, though they seemed positively harassed, cringing close to the ground as if they remembered the head-pincering gales of winter and expected them to strike again at any moment. It was an odd sensation, first looking at the great white ice cap and then turning around to see so much vegetation.

That evening I pondered Eric the Red's color-coding scheme for the North Atlantic. Eric is supposed to have adopted a sales pitch that presented Iceland as icy, white, and awful and Greenland as green and inviting. Being a murderer, perhaps he found that the Madison Avenue style came easily. Yet Iceland in the tenth century did have fourteen more glaciers than it has today, and the southwest coast of Greenland — where, after a trip as far north as Disko Bay, Eric elected to settle — did have fjords ostensibly more suitable for raising stock than the coastal fringes of Eric's volcanic homeland. Also, southwest Greenland gets a nudge from the Greenland Current, an offshoot of the Gulf Stream, and some of its valleys can be strikingly green in summer. If Eric had visited here during what meteorologists call a static depression (a prolonged period of low barometric pressure), he might have thought himself transported to the greenest, loveliest land on earth, a land where photosynthesis goes on twenty-four hours a day and where every petal on every plant seems to grow almost visibly. A land, in short, that speaks eloquently, if seasonally, of Thor's plenty.

Early next morning I took the *Aleqa Ittuk*, a little ten-knot passenger ferry, from Narsarssuaq to Igaliko. We chugged along steep-flanked Tunugdliarfik Fjord toward the blue height of Nunarsarnaq, a mountain shaped like a miniature Matterhorn. The skipper, a Greenlander with a cleft palate and cracked sunglasses, neatly parried solitary chunks of drift ice that had been carried southward from East Greenland glaciers by the cold East Greenland Current, around the horn of Cape Farewell, and into Davis Strait — a long haul, and they'd been properly whittled down by it.

It was this drift ice that bottled up fjord entrances during the so-called Little Ice Age of the fourteenth century and severed the European connection for latter-day Viking colonists. Right now, though, the ice wasn't bad and we made a fairly smooth passage. The only incident of any note during the trip occurred when a Greenlander's drunken girlfriend flung his sneaker overboard. (The victim merely shrugged and flung the other one overboard, too.) After two hours the boat bumped against a little landing place. Two Swiss hikers joined me on shore, reconnoitered their gear (they had enough to outfit a small sporting goods store in Zürich), and trotted off briskly into the mountains.

At two miles the rock-strewn, up-and-down track to Igaliko is one of the longest roads in a country where any road at all is a curiosity. I followed this track to the village through a series of abrupt swivels and switchbacks that made me feel as if I were wandering around inside a vast intestine. Toward the end I began to see the linty forms of grazing sheep, all apparently unaware that this was Greenland, not the plains of Wyoming, and they weren't supposed to be here.

I've often thought the Home Rule government encourages sheep farming in order to keep its constituents from being sucked dry, since the arctic mosquito refuses to live near sheep. Whatever the reason for this aversion (bad odor? wool too thick to penetrate? a preference for more intelligent forms of life?), I've never once felt the tell-tale incision of a mother-to-be while visiting a farm in Greenland. Yet this dearth of mosquitoes may be the lone agricultural virtue of trying to raise sheep on an island whose spectacular winters always seem to kill off about half the stock.

Being the Viking Vatican City, Igaliko had the Viking equivalent of Saint Peter's, a cathedral dedicated to Saint Nicolai, the patron saint of sailors. I walked through a field thick with chamomile and came to all that remained of this cathedral — a few scattered blocks of red sandstone. It had been relatively large, 95 feet by 53 feet, and also rather ornate, with windows of green glaze instead of the usual stretched calves' stomachs. Dust to dust: Now it looked so ruinous you'd be paying it a compliment by calling it a ruin. Over the years the natives had dismantled it block by block for their own significantly more modest red sandstone dwellings. This meant that each of the fifty or so Greenlanders in Igaliko was living in, however shabby, a consecrated dwelling.

I pitched my tent half a mile or so from the village, down by Igaliko Fjord, where I could be watched and evaluated. This was an old Indian practice whose wisdom I value: Never move in on your hosts unless distinctly invited to do so. Strangers are, by definition, untrustworthy — they want your land, your artifacts, or your women. I know I wouldn't take kindly to anyone putting up a tent like mine, part Christo and part Buckminster Fuller, in *my* back yard.

That evening I did receive an invitation, of sorts. I was admiring somebody's lawn, all ragged oil drums and old fish bones, when a ten-year-old kicked a soccer ball in my direction. Soon we'd adjourned to a nearby soccer field, which was just as thick with chamomile as the old Viking churchyard. Possibly I should have made myself some tea on the spot, for chamomile tea is very soothing and might have helped me deal with the bloodthirstiness of Greenlandic soccer.

Before long we were joined by two other kids. In time a match was suggested. Since none of them showed any enthusiasm for being on my side, I ended up playing against all three, a White Man outnumbered by the natives. Right away I realized they were just as inclined to kick me as kick the ball. One of them would block me out, another would bash away at my shins, and the third would drive the ball triumphantly to a goal. Or all three would dedicate themselves to bashing away at my shins, forgetting about the ball. The underlying message seemed to be "Off the White Guy." The great wide world of getting and spending — *things, things, and more things* — may be his, but now we've got him on the holy ground of our soccer field and let's see who's boss. I figured I was getting what they got every day in their Danish-oriented schools, taught by Danes who (for the most part) hate Greenland; I figured this especially when the mockery of a match was over and one of them — he couldn't have been more than eight years old — called me a lousy Dane. I told him I was a lousy American, not a Dane. This seemed to jam his circuits, for in all probability he'd never seen an American except on his TV.

Back at my tent I nursed my wounds and briefly considered a return match, *mano a mano*, wherein I transformed these kids, each in turn, into Inuit gelatin. But as the bright, lustrous evening wore on, I began to shake off my irritation. A trail of opaline clouds crowded the horizon like vistas in a Piranesi drawing. A couple of wayward icebergs rested contemplatively in the fjord, and the fjord itself had turned a rich Mediterranean blue. Once when I looked up, I happened to see a sea eagle poised on magisterial wings above the knurled summit of the mountain behind my tent. It was a scene of peerless tranquility, tossed out in Nature's devil-may-care way, which says: Just open your eyes, my friend, and I'll astonish you every minute of your life.

Next day I awoke to rain as intense as an Asiatic monsoon, except it was much colder and, wind-driven, flung itself around Igaliko in big horizontal sweeps. I've always thought Greenlanders were born waterproof (often you see them sitting outside in snow or rain as if they are sunbathing), but this was too much even for them, and everyone — including the cheerfully demented old man who had stood beside his house and played his concertina all during the previous day — seemed to have retreated indoors. I decided I would use this opportunity to try and locate a guide for my overland trek to Qaqortoq.

Qaqortoq (Greenlandic for Julianehåb) was a six- or seven-day walk from Igaliko and the trip took in several Viking farm sites as well as the old Hvalsøy Church, possibly the last Viking stronghold in Greenland. The route was not without its perils, ranging from lack of paths to suddenly plunging couloirs, but it was probably no more perilous, merely longer, than my trek to Hornbjarg. Yet here — especially here, in this polyglot, puzzle-headed, much-imperialized country — I felt I needed someone to serve as ombudsman between myself and his land, else I'd be violating one of the oldest of all travel rules: The native knows best, and even if he doesn't know best, at least he'll manage to get you into the same trouble he's always gotten into himself.

I visited a man named Gustav Olsen, who possessed a smile as calm as a sheltered inlet. But when I told him where I wanted to go, he quickly swallowed his smile and shook his head, saying he was afraid (*"Niangiannatuk!"*) of the *qivigtoq*. This *qivigtoq* inhabited a little gulch called Itigvitaq only a mile or so from my chosen route. He told me this story:

Fifteen years ago three friends went out to hunt seals in Igaliko Fjord. Each of them had been drinking heavily for several days, case after case of beer, and they had a rather exalted notion of how many seals they'd get. Several hours out of Igaliko they rammed their speedboat full throttle into a submerged ice floe, and the boat capsized. Two of the men drowned, but the third, Anda, reached the shore suffering badly from exposure but even worse from shame. His two companions had died, but he had not died with them. He realized it would be impossible for him to hold up his head in human society anymore, so he elected to live out his days, alone and shame-ridden, in Itigvitaq.

"The fellow doesn't sound so frightening to me," I remarked. "A little sad, but not frightening."

"Would you be frightened if you knew he could change himself into a polar bear? *Ela,* would you?"

"I suppose so . . ."

"*Assâssakâq!* He can. It is well known. Would you like Anda to eat you up, White Man?"

By the end of the afternoon I still hadn't managed to locate a guide from among Igaliko's housebound citizenry. Some were afraid of the *qivigtoq;* others, of unpleasant weather. Some said they'd be too busy tending their sheep; others, their idleness. Hardly anyone wanted to miss Friday's *kaffemik* (coffee orgy), a sort of party where everybody sits around drinking cup after cup of strong coffee until eventually their hearts blast out of their chests like rockets from a launching pad.

At last I met Paulassie Egede, a raw-boned, dimple-cheeked, gap-toothed man in his early thirties. Paulassie lived with his girlfriend Meqqoq in a house whose walls were papered with hundreds of yellow labels from Dole pineapple cans. When I commented on this, he told me he'd been drunk one night and had seen a large shipment of canned pineapple sitting on the dock; he was so impressed by the decorative potential of the brightly colored labels that he tore them all off and covered his walls with them. The canned contents he threw into the fjord, at whose bottom they presumably still lie.

Paulassie was something of a wag, both worldly and a little old-fashioned. He grew cannabis — thin, scraggly, awful-looking plants obviously ill at ease in a Greenlandic window — and claimed he'd be perfectly content to spend his whole life hunting ravens and ptarmigan, soaking up welfare *kroner* from the Danish taxpayer. He asked me where I came from and seemed genuinely disappointed that it wasn't Alabama, because his most cherished possession — apart from a 70 Winchester .270 rifle with a twenty-two-inch barrel — was a University of Alabama sweatshirt he'd bought in a Copenhagen shop specializing in American sweat items. Someday, he said, he hoped to visit Alabama, if only to see what kind of place he was promoting with this shirt.

Paulassie's relaxation in his body bordered on laziness, not a quality you ordinarily look for in a guide. But he did know the mountains, and he had helped some Danes excavate the old Viking farmstead at Sigssardlugtoq last summer. Also, he was not afraid of the *qivigtoq,* nor dedicated to sheep, nor interested in the coffee orgy (he preferred beer orgies).

"*Ela,*" he said with a grin, shaking my hand. "You are Mr. Peary and I will take you to the Pole . . ."

That evening I was brushing my teeth in the rain when a snow bunting dropped down and perched on my tent and gave me the strangest look, as if to say: What in the name of reason *are* you doing, White Man?

July 26. A bright, almost tropical day. As we walk, the Greenland Shield seems to shed more of its thin vegetative garb in favor of naked granite. We skirt the mountain Nujuk, which Paulassie says is unclimbable, owing to scree that always brings your best intentions down with a rubbly flourish, and carry on along a brusque, serpentine ridge. At every col and gully there's a thicket of dwarf arctic birches, dwarf rowans, or dwarf willows, none higher than eighteen inches, through which we have to stomp, a process quite a bit more arduous than boulder hopping.

Every fifteen minutes Paulassie stops and drinks a lingering, salutary draft of stream water. Every half hour he disappears to rustle up a ptarmigan for supper, returning ptarmiganless but — I suspect from the sound of the ricochet — having shot a rock or two. His sneakers have holes in them, as do his sweat socks. Every hour or so he pauses to give both these sets of holes a rest. Sometimes he just pauses and doesn't drink water and doesn't rearrange his footwear. What's up? I inquire. I'm thinking about my girlfriend, he says. She has a very fine *utsuuq*. But he also says his grandmother always told him to proceed slowly in the mountains, since they're full of evil dog-headed creatures called *erqigdlit.* (While *he* wasn't really too worried about these *erqigdlit,* he says his feet sometimes seemed to be quite worried.) Continuing on, we hike past clumps of rosewort and gay rosettes of purple saxifrage with petals flared eagerly toward the sun. More very fine *utsuuqs.*

I hardly mind our leisurely progress, since I'm not in a hurry to catch the 5:17 to Poona Junction, the 6:09 from Grand Central, or the evening Aeroflot to Tashkent. One of the purposes of travel is to avoid your destination at all costs (once you're there, you're there, and you'll never be permitted that long bated breath of anticipation again), and once in a while Paulassie does point out to me the most extraordinary things. For example: We shinny our way up the polished chute of a defunct waterfall, ascend and descend a mountain called Akudlipqaq, and then enter a remote, handsome, seemingly unfathomed valley, when Paulassie stops dead in his tracks and draws my attention to — four Carlsberg lager cans! My own, he says proudly. I came to this place two years ago to hunt ravens. *Sóruna,* I tell him, and you've made it ugly. Beer is not ugly, he protests.

Midafternoon: We take our lunch on a ledge overlooking Igaliko Fjord. I'm gazing down on the fjord, a blue fallen peacock feather, when Paulassie interrupts my reverie to tell me that I'm sitting on some of his ancestors. I peek through the interstices of an apparent pile of rocks to observe three dearticulated human skeletons stacked up inside like logs, with a snow bunting's nest built in one of the rib cages. One of these skeletons is a child's; a few thin wisps of black hair cling to its tiny lemon-colored skull. The other two have patches of green moss growing on them like a close-cropped fuzz of hair (nothing inspires a bryophyte more than the proximity of a good human skull). A family of famine victims, Paulassie says.

Probably starved to death a hundred or so years ago. Are they buried above ground because of upward pressure in the permafrost? I inquire. No, he replies, they're buried above ground so they'll have a beautiful view for all eternity.

Later we pitch our tents beside a cirque of unprecedented turquoise called Tattersuit, which has gullies of snow dotting its shoreline like exclamation points. Tattersuit is an almost impossible lake to read since it has no obvious plant or insect life, just a few wasted silver-gray cattails sticking up like beacons. But in half an hour I've caught eight full-grown arctic char, each seven inches long, with a Dardevle lure. These char have precious little to feed on in Tattersuit, mostly just plankton, but that turns out to be enough for them to continue living even though they hardly grow at all. Paulassie is less pleased with these *erqalut* than with the can of beans he's extracted from my pack. He's never tasted beans before, so he flings the can into our dry willow root fire and PLONK! it soon explodes. The beans are cooked to perfection. Unfortunately he insists on boiling the char for almost an hour, which gives their delicate pink flesh the consistency of an old Goodyear tire.

A cold, lovely evening. Ripples like frost heaves on the lake. Tent rustling softly in the wind. The distant bark of an arctic fox. A few random mosquitoes browsing around our campsite . . . even *they* seem to contribute to the evening's general beneficence, for *Aedes rearcticus* cannot perform its surgery — wing muscles won't function properly — unless the temperature is a minimum of forty-five degrees, which it isn't right now.

"Happy is the camp that doesn't possess mosquitoes," I observe, citing an old Micmac Indian proverb of incontestable wisdom.

"I will tell you a mosquito story, Allaut," says Paulassie. He has taken to calling me Allaut (Pencil) because every time he looks up, I seem to be scribbling something with my no. 2 Ruttledge word processor. Here is his story:

Once upon a time all the mosquitoes in the world lived on a little island in Igaliko Fjord called Qerqertarssuaq. They kept entirely to themselves, never leaving the island, just paddling their kayaks around it occasionally. On the mainland there lived a man named Niliq who made a habit of robbing, stealing, and cheating his friends. The last straw came when Niliq stole a boat full of whale meat. He rowed the boat over to the mosquito island, where he assumed no one would be able to find him. But his friends followed him by the scent of the whale meat. They decided to put an end to Niliq's thievery once and for all, so they beat him to a pulp and left him lying dead on the rocks. The mosquitoes, who had been going through a period of very poor hunting, approached the man and started to drink his blood, whose flavor they liked very much. After that the mosquitoes left the island and forever after hunted people, for once they had tasted human blood, they preferred it above all else.

July 27. Late in the morning we climb down a lateral moraine undulating to the fjord and come upon a lush meadow of dandelions. You can always tell the old Viking farms in Greenland by the presence of this common suburban lawn pest. The land may have been undisturbed for hundreds of years, but dandelions are perennials that respond enthusiastically to the labor of earlier times; the original laborers and their livestock may be dust, but the dandelion still extends its powerful taproot to near-permafrost levels and still raises its bright yellow head in June and July.

At this site, Sigssardlugtoq, Paulassie worked for part of last summer with two Danish archaeologists and one other native hand, disturbing the poorly aerated soil yet more by putting it in a bucket and then shaking it over a sieve. He points to a Viking longhouse, revealed only by its rectangular embankments, and a few scattered stones, either a church or a cow byre, hard to tell now. He says the Danes found evidence of a spring flowing through a stone channel to a basin on the longhouse floor, and then another channel where the overflow could be diverted under a wall. Running water! That was a luxury nobody in Igaliko possessed seven hundred years later . . .

I search around for artifacts and manage to locate an empty potato chip bag, a rusted tin of pilchards, some toilet paper, a scattering of shotgun shells, and another mound of beer cans. Greenlandic kitchen midden, circa 1986. "This time not me," Paulassie declares, pointing to himself and then to the cans (he knows his own litter like he knows the back of his hand). I guess that a couple of seal hunters stopped off here and enjoyed a picnic of potato chips and beer among the warty-fruited dandelions, and I leave their artifacts exactly where I found them. Maybe some future archaeologist intent upon reconstructing the late twentieth century will be grateful for my foresight.

An aside on litter: In the Ecuadorian Amazon, I once followed a trail blaze through the jungle of Shuar Indian homework. Mismanaged sums, botched multiplication tables, and mangled long division, all stuck in liana vines marking the path. The source of this detritus? The kids trek in each morning to get their schooling and each afternoon they happily throw this schooling away. But a few days later, when I passed through again, every last wretched piece of homework had been snatched up by the glutinous Jungle Madre, digested and gone forever. In Greenland, however, it's been estimated that a piece of paper might survive intact, with all the print still legible, for a minimum of five hundred years, whether or not anyone is present to read it. For Dame Permafrost refuses to dine on the nondegradable software, hardware, or any other ware belonging to the modern age. Impossible to digest the stuff. In this she shares the good taste of a Greenlander who doesn't want Carlsberg empties and used toilet tissue — not to mention soggy insulation, unsinkable polystyrene fish boxes, and disemboweled mattresses — lying around his house. Trouble is, she doesn't want that litter in *her* house, either. So it just sits outside in limbo, abandoned and unloved, brooding everlastingly on itself, with a half-life scarcely shorter than uranium.

Back up the same moraine, higher and higher, using rock climber's logic: Take it easy if you have any interest in knowing what ledges, precipices, jams, or cracks the future holds for you. We camp near an unnamed lake whose rocks press together on its clear bottom in a deep hug of silence. Rowans and birches grovel among the lichen, and one immaculate pale yellow arctic poppy peers forth demurely from a sprinkle of glacial till.

Another lucent evening. Ptarmigan, crowberries, and bannock for supper. The bannock (my recipe, tried and true, with wheat germ and a dash of cinnamon) Paulassie proceeds to fling into the air and shoot down like a clay pigeon, declaring it the very worst thing he's ever eaten. Far worse, he says, than the green juices the old people would scoop from a dead walrus's stomach and use as a garnish. I'm honored by the compliment. Finished Epictetus and started a Simenon mystery. Slept so soundly, mummified in my sleeping bag, that I might have been winging my passage along one of the back roads of Eternity.

July 28. On an escarpment whose scoured surfaces give it the candor and toughness of the very old. We leap across a torrential, exhilarating river of melt-off no more than a foot and a half wide, then slide along slabs of granite laid out like enormous planchets of sheet metal. Paulassie pauses to contemplate his girlfriend or shoot something, hard to say which, and I scramble on alone up a deep fretted ledge to get a closer look at a waterfall frayed into dreadlocks by the wind. From the top I can see the snow-streaked mountains north of Nanortalik, seismic skips seventy-five miles away; directly below me, icebergs ride silently in the fjord, like motes of dandruff on a blue serge suit.

Now I catch sight of an arctic fox loping off on its uneven course, a figure in a frieze jauntily come alive. I'm startled. We haven't met with too much sentient life on this trip other than our occasionally sentient selves. Just snow buntings, a few ptarmigan, some errant sheep, the char, and the mosquitoes. It's safe to say that Boston Common on a good afternoon boasts more wildlife. So too do the books of certain writers, wherein the North seems to be a logjam of wildlife — as it can appear if you're traveling from herd to herd in a flying machine. But if you're merely walking or getting a lift with a hunter, you'll see northern landscapes from the very different perspective of your stomach, and then you'll be glad you brought along those desiccated strips of whale jerky or that freeze-dried ersatz.

Once upon a time Greenlanders made their big empty land less lonely by providing every last pawky scrap of its furnishings with a soul. Every animal had a soul, and a hunter needed to pray to that soul directly after killing its bodily component or it would take offense. Plants had souls. Drops of water had souls. Even human beings had souls — two souls, in fact, each no bigger than a medium-sized snow bunting, the first located in the throat, the other set in a bubble of air in the groin. Even laughter and sleep had souls. Thus a person could never travel anywhere without some sort of companionship, nor could he travel anywhere and presume that his medium-sized snow buntings were better than Mr. Nepheline Syenite or Mrs. Gabbro, upon whose bodies he was being permitted to tread. Better than a rock? Better than a warty-fruited dandelion? *Tássaqa!* (Not too likely!)

Up until the year 1721 a Greenlander could converse amiably with his own sleep and maybe even his insomnia, but in that year a Danish pastor named Hans Egede arrived on the island looking for the ice-trapped Viking colony, to whose long-pent-up spiritual needs (he assumed the Vikings had retreated into heathenism) he hoped to minister. Egede didn't find any Vikings, but he did find a group of heathens, so he aimed all his High Church artillery in their direction, daily preaching against the sins of gambling and profiteering, which did not exist in Greenland. He did not win many converts until he started to whip the *angakut*. Then people seemed to appreciate his point of view, though they still wondered what this "daily bread" was they were supposed to be asking for. Egede changed that to "daily seal," yet he made it clear to them that seals didn't have souls. Nothing had a soul unless it first submitted to baptism, and he wasn't about to dunk a ringed seal in holy water. Only human beings were allowed that treatment. One by one Hans Egede baptized these human beings — whereupon Greenlanders got the chance to be just as lonely as everybody else.

July 29. Tonight I'm a little shaky. Can't get to sleep for a change. The events of the day keep coming back to me like a melody I can't thrust from my head. I'm writing these notes by flashlight inside our tent, with a rising wind and Paulassie's deep hoarse snores to punctuate them.

Toward midday we were approaching Itigvitaq. At the last moment, Paulassie wanted to steer clear of Anda the *qivigtoq*'s gulch, all crow's feet and prattling rivulets, but I offered to raise his guide's fee by adding on the return ticket from Qaqortoq to Igaliko if he'd make the detour. He relented, yet he was obviously uncomfortable about it, the first time I'd noticed him ill at ease on the entire trip.

We first spied Anda from a considerable distance, a tiny gnomelike figure squatting in front of a heap of stones. Every few paces Paulassie called out, in traditional fashion, We are friendly, we are friendly, we are friendly. No reply. At last we stood opposite him. He peered up at us with little discernible interest.

Anda's wrinkled parchment skin carried a few stray whiskers — he looked a bit like a Chinese mandarin. He was dressed entirely in sealskins that even the mosquitoes seemed to find a little rancid. "*Qaqqislunga qatangajualaq,*" whispered Paulassie. (He has not wiped off his snot in many years.) From inside the hovel of stones wafted the odor of burnt oil and ancient, petrified sweat.

Slowly his mouth opened to reveal a long tooth extending fanglike from his upper jaw. Then he asked me for tobacco in a voice that sounded cracked at the joints from disuse and bad weather. I offered him some of my Balkan Sobranie, which he grabbed immediately and stuffed into his short-stemmed pipe — the pipe of choice in the North because it allows less time for smoke to condense and it burns like a furnace.

"*Qujanaq*. Perhaps you will give some food to an old man?"

I rummaged in my pack for something he might consider edible, putting aside a few packets of freeze-dried camper's cuisine. Apparently he thought I was giving him these packets, for he grabbed one of them — beef stroganoff — ripped it open, and poured the powdery contents into his mouth, smiling gratefully.

"*Qujanaq*. I have always liked White Man."

Now it was his turn to play host. He crawled into his heap of stones, returning with a thermos of ice-cold tea. Drink, he instructed us. We drank straight from the thermos (he seemed to have no cups) even as I wondered what diseases unknown to modern medical science I might pick up from a man capable of turning himself into a bear. The tea tasted like a decoction of swamp water and acid indigestion, yet custom decreed that we not offend him by spitting it out or getting sick.

We were his first real guests, Anda told us, since two Danish hikers dropped by nearly a year ago. I asked him where he got the tea and he said it came from his son Søren, who visited in the spring and the fall, hauling up supplies of tea, sugar, and flour along with batteries for his tape recorder. Now he inserted a cassette of music by the early 1970s pop group Seqijaq (Good-for-Nothings) into this recorder, and we were treated to a sound not unlike a herd of wildebeest in heat. Evidently he hadn't cleaned the tape heads in a long, long time.

"*Ela*. How many winters do you think I have?"

"Seventy," I replied, trying to be diplomatic.

"I am forty-seven."

All at once he burst into tears and retreated into his hovel, where I could hear him crying with the softness of a child. My diplomacy had failed miserably, and now I felt rather miserable myself. I had not wanted to hurt this man: to be alone for fifteen years in these unnurturing mountains ranked him, in my mind, with Jóhann Pétursson, if not the Desert Saints. It didn't matter that he set new records in the categories of filth and bedragglement or even that he was maybe a little crazy. To me he was genuinely heroic in his pursuit of a fate hardly more supportable than the antlers on the great Irish elk (they weighed the animal down and eventually killed it).

Anda appeared at the doorway with a battered old Lee-Enfield rifle aimed directly at my head.

"Go! Leave me with my shame . . ."

We went. Fifteen minutes later I looked back, and my last glimpse of Anda the *qivigtoq* was of a man older than the hills squatting beside his stones and smoking his pipe in apparent contentment.

July 30. Snow: like tiny parachutes of eiderdown floating from the sky. I'm still feeling somewhat uneasy about yesterday's incident. Also a little scared. The twin adrenal meatballs keep you aloft for a while, but when they have finished with their edifying work you splatter down messily to terra firma. Twenty-four hours later, Anda's stubby brown finger on that trigger seems to have been etched in my mind just a few minutes ago. Only now do I recall that he had some sort of ring — a wedding ring? — a joint or two down from the bluish dirt under his nail.

"You see, Pencil?" says Paulassie. "It is not a good thing to steal another man's solitude."

July 31. Slow descent down a slope seamed with gullies. Then through a thicket of dwarf rowans a foot high and possibly two hundred years old, full of ptarmigan burbling sweet nothings to each other. Finally we come out on a tidal estuary and head west along Hvalsøy Fjord. The island Arpatsivigaq rises up on one side of us like a vast brooding joss-house idol and the mountain Qaqortoq rises up on the other, higher still, the deity to whom the idol is offering homage.

Then we see the old Hvalsøy Church resting in a field of bright buttercups. It's a church after my own heart, being humble (about the size of a medium-sized chicken coop), disused (since the fifteenth century), and ruined (very). Roofless, too. It reminds me of certain ramshackle Irish abbeys whose stonework the local farmers are not beneath shifting to their garden walls. Nowadays God enters such abodes directly through the clear open air rather than through the more difficult and oft-haphazard mediation of an earthly appointee. And when there's not a soul around, as here, He simply readjusts His sights and infiltrates the souls of rocks and the ever-present souls of mosquitoes.

On September 14, 1408, the last Viking marriage of which

any record exists took place at this church. After that ceremony, silence. A haze of last things. The newly married couple, Thorsteinn Olafsson and Sigrid Björnsdottir, stand for a single halcyon moment among the buttercups and then disappear forever. Their descendants disappear forever, too. Atoms have been smashed, artificial life installed, human life spawned in petri dishes, and the Moon walked upon, yet the fate of these last ice-bound Vikings in Greenland remains a mystery. Did the absence of new blood doom them to a slavering idiocy? Were they perhaps kidnaped by Barbary pirates and peddled in the white slave trade? Might they have been wiped out by the Black Death? Did they travel across to North America and, as the explorer Vilhjalmur Stefansson once suggested, mingle their debased genes with the rather better genes of the so-called Copper Eskimos? Or were they simply eaten alive by the mosquitoes? History fails to provide an answer. European history, that is. A little over a century ago, an old Greenlander named Abraham told the Danish folklorist-mineralogist Heinrich Rink the following story, which he'd heard from his father:

In distant time Greenlanders and Vikings lived together on the shores of Hvalsøy Fjord. One day a Greenlander was paddling off to hunt birds when he met a Viking standing near the water. The Viking said: "Silly man! Do you actually think you can hit a bird with one of those little darts? Why, I bet you can't even hit *me* with one of them . . ." The Greenlander obliged him with a dart between the eyes. This killing infuriated the other Vikings, so they raided a Greenlandic camp and killed everybody, including women and children. Now the Greenlanders were infuriated. They paddled their kayaks to Hvalsøy Church, which they set afire. Inside the church the Vikings were tossing the head of a Greenlandic woman back and forth as if it were a ball. So intent were they with this game that they didn't realize the church was on fire. They continued to toss the head back and forth until they burned to death. All except Ungartoq (Ingvar), their leader. He leaped from a window with his young son under his arm, warding off arrows with his bare hands. A Greenlander named Kaissapê pursued him into the mountains, where Ungartoq paused for a moment to fling his son into an icy tarn (better the boy should die by his father's hand than the hand of an enemy). Finally Kaissapê caught up with Ungartoq in the shadow of Igaliko Cathedral. He shot arrow after arrow into the man, but to no apparent effect. Yet he had one last arrow, which came from the lamp rack of his barren wife and was endowed with magical powers. Kaissapê shot this arrow into Ungartoq's heart and thus the last Viking in Greenland fell down dead . . .

Later, while inspecting the church, I notice quite a lot of old char on its worn granitic blocks, as if someone had indeed attempted to torch it long ago. I tell Paulassie old Abraham's story, which he has never heard before. He merely shrugs and says:

"My people and your people, Pencil, they never did get along together."

August 1. Final day. It's warm and windless, which means a diaphanous bonnet of mosquitoes around our heads. They scent out every little morsel of flesh where the Tiger Balm hasn't been applied and then they dig in with an enthusiasm that suggests we might be their last chance for motherhood. Paulassie says the best antidote is beer, lots and lots of beer. I'd heard of smearing seal oil or reindeer fat on your skin, but this was a new one to me.

"You rub *beer* over your body?"

"No, you stupid *qalunaaq*. You drink it and then you will forget all about the mosquitoes."

At the lake Olserssuaq we observe our first human beings in almost a week not begrimed with filth or supernaturally endowed. And what human beings! They're a bunch of young girls cavorting naked in the lake, giggling, laughing, splashing, and sometimes even swimming in the upper six inches of water (below that is hypothermia country). Paulassie and I glance at each other and nod appreciatively. After so many rough albeit rewarding rocks, so many undulating moraines, so many yellowing bones, these girls are like a vision of earthly delights visited upon the last place on earth you'd expect to find it — a secluded glacial valley with an icy glacial lake and a few ground-hugging white rhododendrons which seldom bloom for more than two weeks a year and are blooming right now . . .

Whereas most towns in Greenland seem to represent urbanization on the South Bronx model, Qaqortoq (pop. 1,700) represents the Danish provincial model. Perhaps that's because it has a fairly large population of provincial Danes to keep it free, more or less, from Greenlandic urban blight. ("Qaqortoq has no recreational homicide," a local friend of mine once quipped.) The town is quaint, safe, and pleasantly dull. It even has Greenland's only village green. All in all, it seemed like a perfect place to rest my weary limbs after our long trek. But shortly after we arrived, I dropped by the shipping office and learned that the MS *Disko* would be docking here tomorrow en route to Nuuq and wouldn't be coming back for another two weeks. If I didn't take it, I'd be stuck in Qaqortoq for those two weeks, which seemed like rather a long time to spend staring at Greenland's only public fountain.

I bought Paulassie, a bit belatedly, a good pair of hiking boots and treated him to a whale steak at the Nanoq Restaurant.

"Where are you going now?" he asked me. Tomorrow he'd be heading back on the *Aleqa Ittuk* to his Dole pineapple-labeled house in Igaliko.

"To Nuuq. Your capital. The Western Settlement of the Vikings."

"It is very bad in Nuuq. They say it has the most suicides of any city in the world."

"I've heard the same thing said about Aasiaat."

"Aasiaat is not a city . . ."

After dinner I walked him to a dance that was being held in the local recreation hall. We promised to stay in touch, but I knew we wouldn't. The wilds exalt camaraderie only when you're passing through them, and it's a far more intimate camaraderie — no bodily odors barred, nothing sacred, and virtually nothing private — than the fellowship of the city. But we weren't in the wilds anymore and even now I could feel my friendship with Paulassie ebbing gracefully. His mind was already on the girls he'd be meeting at the dance; my mind was on the boat trip up the coast. He picked up a package of condoms at the door (condom machines were no good; they froze), turned to me and said, *"Ingerlalluariarnat, Allaut!"* (Safe journey, Pencil), then walked in. He did not look back.

Before I turned in, I watched a hunter flensing a ringed seal down by the dock. The animal's dead eyes stared up at me like the eyes of a Mexican child painted on velvet. Its flippers lay at its side like futile baby hands. The hunter was just beginning. First he took out the gall bladder so it wouldn't explode and taint the meat; then he removed the liver, a culinary delight of the first order, especially when it's raw and still steaming; then he reached in for the intestines. While his countrymen were being trained by the Danes for the high-tech future (or the sheep-farming nonfuture), here was this man seated happily in a puddle of gore, with the startlingly thin intestines of a seal like a cat's cradle in his hands. He wiped his greasy fingers on his anorak as if he were trying to waterproof it.

We chatted. The man spoke without looking up, in rhythm with his work, as if each word were a sinew. He told me this story:

Once a raven and a seagull got into a fight over a piece of

meat. The raven was on the Inuit's side, the seagull on the side of White Man. They fought for days, for weeks, even for months. Whoever won, his side would be the stronger. They tore and bit furiously at each other. At last the seagull won: White Man would be stronger and more plentiful than the Inuit. But by the time he flew away with his piece of meat, it had become quite rancid.

The man told me this with such quiet conviction that I knew it was the truth.

Chapter 10

CITY OF THE DREADFUL NIGHT

Next morning I headed north. Just outside Qaqortoq the *Disko* encountered drift ice, most of which she could breeze through, knocking it aside with a cavalier nudge of her red bow (Greenland boats are painted red or orange to distinguish them from the ice through which they must plod), but sometimes a louder, more resounding *thump* would indicate that a certain chunk of ice wasn't pleased about giving up its berth to this man-made interloper. It had not floated all the way from East Greenland to this crucial spot just to be brushed aside by some opportunistic slab of metal. At this point the boat would reverse and try again, pitting itself against the ice's hidden bulk. Usually the ice would spin sluggishly away, but once in a while it wouldn't and then there was no choice but to go around it. In certain years a field of ice can stop the *Disko* dead in her wake or send her nudging and shouldering up the coast at quarter speed.

As we sailed on, icebergs hunched in the water like a hierarchy of Chinese pagodas, Etruscan palaces, grain elevators, Boeing aircraft hangars, and alpine peaks. Some had been whipped by the wind into castellated spires, others had grottoes and natural arches, and still others looked like frozen shipwrecks of long ago, their masts petrified by whiteness. In a heavy overcast they had an eerie bluish white phosphorescence, but when the sun's slanting beams struck them they'd turn from white to blue to pink and even to black. I saw the Cathedral of Chartres painted a garish violet shortly after we pulled out of Narssaq, a small village where the physicist Niels Bohr discovered, greatly to the consternation of the locals, a lode of uranium ore.

Every once in a while there would be a muffled crack, like a distant fusillade of rifle fire, and a berg would turn turtle. The soapy sheen of its underbelly would rise up to replace its former rough edges and near-right angles, soon to be whirled into a roundness themselves by the potter's wheel of the sea. Or the explosion would mean a berg had calved, flinging out a progeny of brash and smaller bergs to embrace survival on the high seas alone.

If an iceberg were just ice, it would not be able to keep itself afloat when the cold heart of its mother glacier touched the comparatively warm sea. But the packed snow, granular in form, contains millions of tiny compressed air bubbles, which keep it buoyed up. The air in these bubbles might have been breathed by woolly mammoths and woolly men alike during the last Ice Age. Once, when the *Disko* drew close to a berg, one of the crew reached out with an axe and hacked off a piece of ice for the kitchen. Later, in my drink, this ice fizzled and detonated: air being liberated for the first time in maybe fifty thousand years.

The first evening out of Qaqortoq I stayed on deck and gazed at these bergs resting in the soft ectoplasm of their own fog. With a team of porters to bring me food and drink, I could have gazed on them for the greater part of my life, delighting in their unfailing indeterminacy, their lack of structural principle — qualities found only in the natural, not the plastic, arts. Here in Davis Strait you never come close to seeing the same icy creation twice, nor do you ever know what — the Cathedral of Chartres? Annapurna? the Prudential Life trademark? — will put in an appearance next. Around two-thirty A.M. I saw the State Capitol in Montpelier, Vermont, float by with a colony of kittiwakes gabbling on its dome (lots of terrestrial nutrients in that icy edifice) as if they had found the most luxuriant of cliff faces.

Sometimes, when a skipper is more interested in seals than in his destination, a passenger boat in Greenland will sail off on an impromptu hunting expedition. If that happens, the boat may arrive as much as half a day behind schedule, with the deck resembling an abattoir. But the *Disko*'s skipper was a Dane named Torbjørn. Torbjørn was safe, reliable, and courteous, just like his Greyhound counterpart, and we reached Nuuq in quite good time. Two and a half days: twice as fast as a six-oared Viking craft rowing from the Eastern Settlement, once it became overcrowded, to less cramped quarters in the Western Settlement.

Nuuq is located on a peninsula of gneiss and boggy hollows shaped on the map like a wizened fetus. In 1728 Hans Egede moved his mission from the raw and windswept Habets Ø (Hope Island) to this more or less protected place, which he called Godthåb (Good Hope). The mission flourished and became a town; since Home Rule it has gone by its Greenlandic name. Nuuq (The Hood) refers to a saw-toothed mountain just north of town that is supposed to look like a head peering from an anorak hood. I've never met anyone who thought the mountain, Sermitsiaq, looked anything like that. To me it looks more like a neck that has recently undergone decapitation.

The fjords behind Nuuq sheltered the Western Settlement of the Vikings, which made Nuuq a natural stopover on my Viking journey. But even if it hadn't been natural, I felt the time had come for me to observe a bit of urban squalor, northern style.

Southern squalor you expect — malarial ports and choked slums and rampant rivers of cess. But when you head north, you expect a certain purity if only because fewer people would seem to indicate smaller messes. Dumb idealism! Your first view of Nuuq consists of twenty-two identical blocks of project apartments, each five hundred feet long, which taken together house 15 percent of Greenland's total population. Your binoculars will be greeted by a myriad of windows broken by such projectiles as heads and fists (the broken window is a Greenlander's idea of interior and exterior decoration simultaneously). Aim those binoculars a little lower and you might see a clutch of eight-year-olds knackered out of their gourds, sitting and playing in lagoons of greenish sludge. Maybe you'll even see a few old women born in the Stone Age, cases of Carlsberg cradled to their breasts, struggling uphill to the projects. In such a place the venerable image of the Eskimo hunter, harpoon in hand, is no less exotic than a nude calypso dancer.

After I checked into the Seaman's Home — another cheap temperance hostel, with hymn singing in the evenings — I took a stroll around town. As usual, my peregrinations took me past an elephant's graveyard of litter: soggy insulation, holed glass-fiber boats, broken-down prams, old tires, sugar bags for infants, Carlsberg empties, and a sea of flip-tops. After seeing this same landscaping in every port and anchorage in Greenland, the traveler begins to yearn for something different, such as the carcass of a personal computer or a VCR. But no one's throwing away VCRs in Nuuq, not as long as they can pop into a video *pissiarfiq* and choose for their evening's entertainment American films with titles like *The Revenge of the Termite People, Pistol-Packin' Mamas, Teenagers on Mars,* and *Hollywood Wrestling Women versus the Aztec Mummy,* all dubbed in Danish (I admit to a curiosity about that Aztec mummy talking Danish myself).

Near one of the project hulks I overheard a French girl asking a Greenlandic woman in English, a language neither of them seemed to understand, where she could buy a box of tampons. Her pronunciation made it sound like she was asking for a box of tom-toms. Finally, unable to tolerate their mutual confusion any longer, I doubled back, excused myself for being so bold, and suggested to the girl that she try the Royal Greenlandic Trading Post, just down the road.

The French have always struck me as an incredibly fragile race of people, with tiny birdlike bones and ornithic faces. This girl herself was as delicate as a sparrow, but appearances can be deceptive and as I walked her to the trading post, she told me a little about her travels. She had started out in June and had threaded her way up the coast as far north as Upernavik at latitude 73°N. I couldn't begin to guess what the citizens of that remote hunting community thought upon seeing this red-haired wisp of a girl suddenly appear with a forty-pound load on her back. What *did* they think? "They asked me to be their nurse, Monsieur," she said, "but I am a telephone repair-woman, not a nurse . . ." In Uummannaq she had climbed to the top of the heart-shaped mountain from which the town gets its name; I'd heard of this ascent being performed only twice before, once by a Swiss alpinist who had planted the Swiss flag on its summit and the second time by a Greenlander who had tromped up in his wellingtons to remove this flag. She mentioned the difficulty of getting tent pegs into an ice-hard layer of permafrost, but she mentioned it as if it were one of life's chief pleasures.

I asked her what she thought of Nuuq.

"*Merde!*" she exclaimed. The town depressed her so much she was taking the late afternoon flight to Ilulissat.

After I left the girl, I went down to the open-air market Braedtet to buy my supper. As I looked over briskets of reindeer, sides of seal, dead kittiwakes, and fish with eyes like computer buttons, I tried to ignore the two old men who were facing each other, drinking *imiaq* (home brew) and then vomiting it up, drinking again and vomiting again, like a pair of Roman generals at a banquet. This was not behavior calculated to improve my appetite. Soon a woman with a soft round face and shoulders like a fullback's was trying to interest me in what she called, simply, *qimmeq* (dog). Another woman tried to sell me a couple of decomposing seal heads — *mikiaq,* a local specialty. Neither of these items appealed to me much, nor did I have any interest in the aging kittiwakes one man tried to sell me, claiming he'd just shot them this morning. I turned my attention to the fish and rummaged through a pile of halibut stacked up like warped cardboard, found them a little too "high" for my taste, and at last settled on a humble cod.

The person who sold me this cod was an eagerly voluble man whose corrugated face was an entire geography unto itself. His name was Qalaseq (Belly Button). Why he'd been called that he never had the opportunity to learn, because both his parents died of tuberculosis when he was an infant. He was raised by an aunt, who died of tuberculosis when he was eleven. Then he was passed on to his uncle, who died shortly thereafter in a kayaking accident. He was then thirteen years old and entirely on his own.

"That's a lot of deaths for one small child," I said.

"*Ila nâgga,*" he replied. "Not really. Look how many people didn't die. Many thousands. You must think about it that way, my friend, or you will die before your time yourself."

Qalaseq wasn't quite so philosophical about the day he and his wife Pauline were sandbagged out of their village Kangeraatisaaq by the government. He'd loved Kangeraatisaaq, he told me. There he had owned the craftiest team of dogs ever to be tied to a man's traces. There the sea was so full of game that you could close your eyes and your harpoon would strike a seal. There the sky got so blue, it would take your breath away. Like nearly everyone else, he detested Nuuq.

Kangeraatisaaq was one of dozens of small villages along the coast which the notorious G60 Policy in the 1950s and 1960s had rendered obsolete. It was too difficult to provide these villages with the services they hadn't asked for in the first place, and too much of a drain on the Danish taxpayer to keep them afloat even though they'd already been afloat, without Danish *kroner,* for centuries. Besides, joked the Danes, their names were too hard to pronounce. Better to move everyone off to shorter-named places. Thus the refugees were removed from their turf-and-stone houses, shipped off to towns like Nuuq, and installed in apartments with flowered wallpaper and parquet floors. This may seem like altruism, and in a way it was. Yet the Danes had another motive too: they needed bodies for the burgeoning fishing industry. Unfortunately the water temperature dropped a few degrees in the late 1960s and most of the cod — formerly so thick in Davis Strait that you could jig them with a bent pin — decided to move farther south. At least *they* weren't going to stick it out in alien circumstances.

In the fashion of his people, always hospitable even if they hate you, Qalaseq invited me back to his apartment. He lived in Block P — the longest, most monolithic block in town, with nearly a thousand inhabitants. Since the Danes had blasted the foundation from solid bedrock, they couldn't afford to be dainty, so they built Block P and its companion blocks high, long, and compact, like the resettlement blocks in Hong Kong. Twenty years later not a single one of these blocks has been given a name.

We walked the six flights of stairs to the top floor and I noticed graffiti spray-painted on the walls everywhere. Most of it celebrated very fine *utsuuqs* or the Inuit Independence Party, which seeks a complete break from Denmark, but there were also some curious localisms, like: "Pavia is a bearded seal's foreskin." Qalaseq said his neighbor Pavia had been born with a bluish caul over his head, which everyone thought hysterically funny.

One thing about Greenlanders: they tend to find misfortune amusing. I once saw a man return from Denmark in a wheelchair, and when his family met him, they slapped their knees and rolled in the snow, pointing and laughing at the old man (he laughed with them) stuck in this odd-looking chair of metal. In *The Last Kings of Thule,* my favorite book about Greenland, Jean Malaurie describes how the good people of Thule always used to mimic a lame man named Asarpannguaq trying to make love. Cruel, yes, but it's cruelty that serves, or once served, a useful purpose: you've got to be tough in this vale of misfortune or you'll exchange your breath for a pile of stones. There's a saying that Danes beat their children but not their dogs, while Greenlanders beat their dogs but not their children. It's probably true; not once have I seen a Greenlander strike a child. But he will ridicule that child unmercifully or perhaps give him a nickname like Usukitat (Little No-Good Penis) that will stay with him all his life. In Igateq, East Greenland, I once met a hunter named Itiktarniq (Liquid Dog Shit), who was as tough as nails.

Qalaseq and his wife Pauline lived in a single room on the top floor. The room was a combination kitchen, sitting room, bedroom, workroom, and slaughterhouse. A dead seal lay in the sink, its belly split down the middle and white fat peering out. Fish were draped to dry on the balcony. Qalaseq's hunting rifles stood in a corner along with umbrellas and Pauline's cane. The wall had the Oberammergau Jesus side by side with an apparently nonfunctioning cuckoo clock. The floor was, of course, parquet.

"We have lived in this tiny little place since 1968," Qalaseq told me.

"And we will surely die here, too," said Pauline.

Pauline could have modeled for an Eskimo soapstone carving. Short and squat, she wore her hair in a tight topknot, and her cheekbones were so high they seemed to threaten her eyes. She had grown up in Igdlorssuit, she said. As a little girl, she'd sat on the knees of Rockwell Kent, the American communist artist who took a few years off from being hounded by the State Department to homestead in Igdlorssuit. Pauline was a first cousin of "Mr. Kinte's" girlfriend-housekeeper Salamina Fleischer. "Mr. Kinte" himself was the first bald man she had ever seen; her own people always went to their graves with full, black heads of hair.

I was, Pauline said, her second American.

An admission such as this serves as a reminder that Nuuq is still somewhat cut off from the rest of the world. The only road out of town — at seven miles, the longest in the country — leads to the airport, and you can easily take a taxi there every day of the week and still not get away if the weather is bad. If the weather is good and you've had enough of Nuuq, the only place outside of Greenland to which you can fly directly is the Canadian town of Frobisher Bay (Iqaluit), which is almost as dreadful an attempt to urbanize the North as Nuuq itself.

Now Pauline put an enormous plate of *mataq* on the table. *Mataq* is the black-and-white mottled skin and first layer of blubber of the narwhal. Unless eaten alfresco, hacked from the recently killed whale by hordes of knife-wielding enthusiasts, it is usually diced, allowed to chill, and then served raw, often with sugar and berries. I prefer mine neat, since I don't think the flavor — similar to very tasty rubber, with a trace of nuts and oysters — benefits much from additives. Indeed, it seems to lose some of its preternatural starkness, its intimation of cold centuries in the sea, by being dandied up like a pastry.

The Tibetan proverb "To eat lama's food requires jaws of iron" applies to this very un-laman food, too. You chew and chew. The *mataq* yields itself to you slowly, almost unwillingly, as if it were making up its mind about whether or not to let you swallow it. Then you chew some more, if only because you feel the need to prove yourself worthy of this indomitable food.

"You like it?" Pauline asked.

"Better than dog," I admitted.

While we chewed and chewed some more, we watched the TV weather report of three days ago. Nuuq's weather comes on video cassettes specially prepared in Denmark; by the time they get here, the sunny skies forecast by the Danish meteorologist always seem to have turned to a leaden overcast. Certain enterprising people, tired of waiting, have set up satellite dishes so they can get the weather the very same day — though usually it's the weather in, say, Italy. I noticed that Qalaseq and Pauline had arrived at a partial solution to this problem by owning a TV that showed snow general over its screen at all times.

Later we heard yelling and screaming through Block P's tinsel walls. Qalaseq said these fighting neighbors were a couple from Nuusuaq, more casualties of the G60 Policy, another small place condemned to dust. They destroyed each other every night, but a few hours later you'd see them with eyes blackened, faces bruised and bleeding, clothes ripped, grinning from ear to ear and holding hands as if nothing had happened. This was love, Inuit style, I thought: Once upon a time people rubbed noses, now they break them. But you can't remove Real People from the wide open spaces of the Greenland Shield, stick them in a box, and then expect them to be genteel.

During my stay in Nuuq I always seemed to be dodging the only fleet of buses in Greenland, which careered around the streets like mad elephants, dispersing everything in their paths. One day I decided to take one of these buses to a berry-rich hillside I'd been told about just north of the projects. The driver, a Greenlander, had the nervous indecision of a passenger suddenly left alone at the controls of an airplane. He kept playing with the pedals and fidgeting with the switches and glancing over his shoulder as if he expected to be found out at any moment. Once he nearly flattened an old woman who was pulling along a few crates of beer in a wagon, and a minute later he did manage to flatten a half-flensed seal somebody had either lost or discarded on the road. Safe: no. Courteous: possibly. Reliable: certainly not. I got off the bus and walked the rest of the distance to the berry fields.

Another day I walked out of town to do a bit of climbing in the mountains behind the airport. I scrambled up and down slopes that contained some of the oldest rocks in the world, isotope-dated at 3,800 billion years, remnants, so the geological rumor goes, of the earth's earliest terrestrial crust. They were battered, eroded, and browbeaten beyond belief, as if the marathon times they'd seen had pained them physically. Farther and farther I trekked, swatting at mosquitoes and hopping boulders. But even ten miles from town, I could still hear the noisy browsing of bulldozers in Nuuq. In the summer, with almost continual daylight, the busy beaver Danes work overtime day after day, raking in enough *kroner* so they can buy that cute little *pied-à-terre* outside Copenhagen and never have to grit their teeth in Greenland again. I kept thinking that I would be out of hearing distance as soon as I reached the next ridge or promontory, but I never was. However, I did climb so far that I had to spend the night in the mountains, curled up among some very youthful-looking basalts. Around midnight a reindeer came by to chew at a stunted patch of heather and gave a loud snort of astonishment when it saw me — I jerked awake — and then ran off, scared out of its socks (I was scared out of mine, too).

Then one evening I decided to visit Christinemut, Nuuq's ever-popular drinking and brawling establishment, a place of such rowdiness that it makes a South Boston Irish bar seem like a gentleman's hairdressing salon. In Christinemut sooner or later somebody's fist will find the shortest possible route to your jawbone. Indeed, it is often said that only the host of broken jaws suffered here stands between the local dentists and the breadline. I had been warned off Christinemut by Qalaseq, but I figured that if I kept my nose clean and sat quietly nursing a beer in a corner, I would come away with an eyewitness account of a war zone.

I arrived shortly before ten P.M., easing into a densely packed room. It soon became clear that my face was the only white face present, apart from two or three Danes who were so far gone they didn't count as faces anymore. Initially I felt the same sort of paranoia a black man might feel if he showed up by mistake at a John Birch Society picnic. I imagined that these people, hearing some previously agreed-upon signal, might put down their beer bottles and vent their ancient hostilities on me. I had never felt this way in Greenland before, though I once spent a month in the East among people who still regarded whites as the stuff of legend, the half-human progeny of a woman from their district who had coupled with her dog.

But no one paid me the slightest attention. Nor were they paying any attention to the two very fat, very drunk women standing in the middle of the dance floor and trying to knock the daylights out of each other. One of these women I recognized from Braedtet; she sold the rotting seal heads.

"Why are they fighting?" I asked at the bar.

"They are fighting over a man. The one who goes to the hospital, she loses him."

At last I found a table occupied by three men who seemed to represent the only sobriety in the house. I greeted them in Greenlandic, but it turned out they were Japanese, not Greenlanders. They went out with local trawlers to handpick prawns for Tokyo's finest restaurants. I asked them how they liked Nuuq and they said they liked it very much, but as near as I could tell their affection was based on the seemingly endless supply of young women available to them. One of them had even managed to form a liaison with the nurse who'd given him a penicillin shot for the gonorrhea he'd picked up from another nurse.

Soon we were joined by a man who spent fifteen minutes trying to focus his eyes on me, aiming at my face and usually getting only my lower neck. Between swigs of beer, he would swing his head from side to side and say: *"Silagiq!"* (It's a beautiful day.) I translated for the Japanese, who seemed to think this a perfect example of the irreconcilability of cultures, since the day had been solidly gray with a strong north wind. But then another man sat down at our table and our drunk friend said the same thing to him, for which he got a whack in the face, along with this request: *"Baarit!"* (Go fuck yourself.) Just because the poor man had been wrong about the weather? I considered this interchange and suddenly it dawned on me that what he'd been saying, instead of *"Silagiq,"* was *"Silayiq,"* which means "You have nothing in your head." In Greenland a mere rushed consonant can mean the difference between good weather and a trivial mind.

"We think Eskimos rike to drink too much," grinned one of the Japanese.

"You'd drink like that too if you had their lousy rate of exchange," I said.

Later, as I stood at the bar ordering another beer, a woman of middle years who was wearing a loud Hawaiian blouse staggered over to me and said: "*Tingasa*. Let's do it." It would have been absurd for her to suggest that we sleep together, for "sleeping together," in her language of half-swallowed gutturals, clucks, and uvular fricatives would have called up an image of two people lying side by side in bed, lost in dreamland. It was absurd anyway. The woman looked like she'd had more cases of VD than hot lunches.

"I'm too tired," I told her.

"Let's go back and do it," she repeated.

Then a man whom I took to be her husband approached us. He had a low-slung stocky body and quite powerful arms. He moved not with a drunken stagger but in a stealthy crouch, like a beast of prey. A puckered scar zigzagged across his face and pulled the side of his mouth into a conspicuous leer.

"Why do you not want to do it with her?" the man inquired with a scowl.

"I have a bad penis," I told him. That, obviously, wasn't enough to get me off the hook, since all the best men in Greenland had bad penises (gonorrhea). Meanwhile the woman was rubbing her body against mine and blithely tweaking my nose. "Give me a break," I protested, trying to back away.

"You and she must do it," ordered the man.

"*Ela!* You and me must do it," repeated the woman, making a grab for my crotch.

Eight years ago the hunter in Igateq had offered me his wife with the grace of a man doing me an enormous favor. Here I was being offered a slab of meat, nay, forcibly fed it, by a man who had the grace of an abscess. And I would have no part of it.

"Nâgga, nâgga," I said, backing away.

But the man did not seem inclined to let me escape so easily. As I shouldered aside half-stupefied, tottering drunks, he did the same. As I staked out my passage for the door, he moved toward the door, too. He kept his gaze fixed on me like a silkworm moth, which can see, of all the creatures in the universe, only its mate — except I wasn't this man's mate but his intended victim. If I could only reach the door, I figured, I would make a run for it and probably beat him to the hymn-singing confines of the Seaman's Home, where he wouldn't dare trespass for fear of revenge by the Lutheran god. At one point he had me by the sweater, but I pulled away and he was left pawing the air. At another, he made a lunge for me and missed. One moment he yelled that he was going to cut my throat, and the next he made a gesture with the index finger of one hand and the thumb and forefinger of the other that indicated he still wanted me to fornicate with his wife. Did he yearn to be a murderer or a cuckold? It was an age-old dilemma. As for myself, I only wanted to get back and sing some of those delightfully boring hymns.

The man's voice was a fist: "*You and she must* —"

All at once a total stranger pushed roughly in front of me and gave him a stiff uppercut to the jaw. At first the man acknowledged this blow with only a slight blinking of his eyes, as if he had fully expected it (this being, after all, Christinemut, haven for broken jaws). Then he lurched back four or five steps and collapsed on the dance floor, where a couple doing the Attila the Hun two-step danced all over him. He did not seem to take any notice of them at all.

The man who interceded on my behalf had the same low-slung stocky body and the same powerful arms as my nemesis. *His* scar went along his right cheek and then took an abrupt eastward turn and moved jaggedly along his left eyebrow. He looked like the sort of person who'd roll me if I tried to thank him, so I left him alone. In any event he was already at the bar ordering the case of beer that most patrons order at last call to see them through the watches of the night.

"Are you ready to do it now?" I heard the woman yell after me in a curiously hopeful voice as I stepped out the door. Ten minutes later, in front of the Seaman's Home, I was approached by a girl who said: "*Asavakkit.* I love you. Do you have a nice penis?" This girl was quite pretty, but she was only about ten years old and very, very drunk . . .

Enough! If I had learned one lesson from the wayfaring Vikings, it was this: Move on, friend, once the walls grow too thin, the bodies too thick, or the compost too monumental.

What the Vikings would have made of Nuuq, I couldn't begin to imagine. Probably nothing; they would have pulled up their stakes at the first sign of trouble (a surveyor's theodolite? a concrete Mixmaster? a parquet floor?) and lit out for the Territory again.

So it was that I boarded a Dash 7 aircraft for Søndre Strømfjord the very next day. It was an interesting takeoff; the "Rygning Forbudt" (No Smoking) sign was on and everyone was smoking, including the pilot. Not wanting to feel left out, I lit my pipe. After a while the bluish haze inside the plane matched the thick clouds outside it. I took this opportunity to consider the last few days. From a smoky chamber 5,000 feet above the ground, Nuuq did not seem like the worst place in the world anymore, especially when I'd been able to escape from it with life and limbs intact, and with a fat bag of crowberries. Not the worst place at all. That dubious honor was still a toss-up between the Pakistani towns of Darra and Londi Khotal, where everyone from age eight up carries a Kalishnikov rifle as casually as the French carry loaves of bread. Technology kills. Fists merely help the dentist purchase his *pied-à-terre* outside of Copenhagen.

Chapter 11

TRAVELS IN THE LATE STONE AGE

THE EXPLORER SVEN HEDIN is supposed to have burst into tears the first time he saw the rousing snowy peaks of the Himalayas. I might have burst into tears myself the first time I saw the mountains of East Greenland, except that my aircraft was being tossed about by blustery weather and my stomach was where my tear ducts ought to have been.

That was several years ago. I had been the lone passenger on a Mitsubishi turboprop flown by an Icelandic bush pilot friend who was visiting his Greenlandic girlfriend in Kulusuk. I had joined him to see members of the Family of Man who had been unvisited by whites until 1884, when Lieutenant Gustav Holm of the Danish Navy rounded Cape Farewell in an *umiaq,* a woman's boat, and headed northeast into unknown regions. Lieutenant Holm was searching for the lost Viking colony; even at this late date Scandinavians, ever optimistic about their race's survival powers, expected a contingent of Vikings to be lingering in Greenland's remote fjord country. Like nearly everyone else in those days, Holm believed that the old Eastern Settlement was on the east coast, just as the Western Settlement was on the west coast. Both were on the west coast of the island. But until his visit the stately procession of pack ice had prevented Europeans who had sailed in the vicinity — Martin Frobisher, Henry Hudson, William Scoresby the Younger, and numerous whaling crews — from realizing that the east coast was not inhabited by pale-faced facsimiles of themselves but by a tiny group of Stone Age hunters wholly cut off from their west coast brethren. Thus Lieutenant Holm discovered the Tunumiut (literally, "People of the Back Side" — the back side of Greenland), and they discovered him.

The bush pilot's paramour peed unabashedly into a gas-oil barrel in front of visitors and still believed in the inherited lore of her people. (For instance, if you're pregnant and eat seal intestines, your child will be shaped like an intestine.) I thought my friend very lucky to have found such a woman, especially in such a forlorn backwater as Kulusuk. The village was set on an island whose only distinguishing feature was the giant reflector globe of a Distant Early Warning (DEW) Station, one of sixty-one interlocking radar sites spread across the North and designed to provide advance notice of Soviet shenanigans. This globe peered balefully eastward like the lone eye of an intergalactic insect. All else was rock.

Soon after arriving, I went into the village. A light snow was falling. Through its veil I took pictures of dog sleds parked like cars in front of every house. I took pictures of the dogs themselves, squarely built, with duplicitous wolfish glints in their eyes. I took pictures of kids eating Eskimo Pies (a local favorite, called by that very name). Then, as I backed up to get a panoramic view of the village, I lost my footing and slipped head over heels down a steep icy incline, crashing into the cold November sea. It was the hardest, most violent cold I have ever felt and it struck me like a flying brick.

Somehow I staggered up onto dry land, where I was retrieved by a couple of teenage boys and brought to the Kulusuk infirmary. I was shivering furiously. Shock wave after shock wave shook my body from head to toe and back again. The resident nurse, trained in Copenhagen, couldn't stop this shivering. My pilot friend gave me a bone-crunching massage, but that didn't stop it, either. Nor did it benefit from the teams of local kids who came to giggle at this strange palpitating *qalunaaq*. At last an old woman who walked like a duck full of eggs came in and instructed me to take off the rest of my clothing. Then into my exposed skin she began rubbing a very sticky substance that looked and smelled familiar. Walrus oil? Seal grease? Bear fat? No: Crisco. The old woman said she always used it on her sons if they suffered from too much exposure. It shut in body warmth even as it shut out the cold, closing up skin pores like a package. And in a remarkably short time, I *did* stop shivering. The traditional remedy of Crisco had triumphed.

Nowadays whenever I eat a cookie that's moist and chewy or a cake that's high and light I recall the frigid seas of East Greenland and my own mortality.

After being fogged in for a day at the Søndre Strømfjord airport, I continued on to Kulusuk. Kulusuk wasn't my destination even though I rather wanted to stick around and prove to the locals that I was capable of something other than falling into their fjord. This time I hoped to reach the tiny outport of Sermiligaq, which was to be my last Viking destination in Greenland. To reach Sermiligaq from Nuuq, where I'd started out the day before, I had to fly to Søndre Strømfjord, fly from there to Kulusuk, take a Bell 212 helicopter from Kulusuk to Angmagssalik, and then hope that some mode of transportation would be heading up the coast in the next few weeks. Needless to say, Sermiligaq has a reputation, even in Greenland, for being heroically isolated. But in a global village, where you can jet down to New Guinea almost as fast as you can say Michael Rockefeller, difficulty of access may be the last remaining grace.

As it happened, the Angmagssalik helicopter was undergoing repairs and I had an afternoon to kill in Kulusuk. I trekked into the village along a crevice that shimmered with billions of grains of year-round snow. This crevice turned into Kulusuk's main artery, a track of gravel, silt, and dog shit. At last I saw the steep-roofed, small-windowed, corrugated iron shacks I had seen before, some on stilts to permit more storage space, others in need of stilts to prevent collapse.

I made a few inquiries about my former friends. The old Crisco woman was dead and so was the man who'd told me stories in the infirmary like the one, apparently true, about his grandfather wanting a new wife so badly that he ate the old one. At that infirmary I encountered an extremely officious new nurse who looked as if she'd sooner try radiation therapy than let one of her patients be smeared with Crisco. A young man came up to me and said: "*Ela!* I remember you. But I don't remember how I remember you." We talked for a while, and I realized that he'd been one of the kids who had thought my shivering so funny. Now he had two kids of his own, along with the job (he proudly told me) of scrubbing the plastic flowers and wreaths at the local cemetery.

I was walking back to the airport when a man who was the perfect likeness of Nanook of the North intercepted me. Across a gap of 10,000 years, he said: "*Ela-ela, qalunaaq*. I need beer. Give me some *kroner* or I will die."

I gave this man his *kroner*.

Later I took the twenty-mile helicopter shuttle to Angmagssalik (pop. 1,100), a town draped over a few gravelly hills with not even a handkerchief of flatness for a fixed-wing aircraft.

Gravelly hills; a bad harbor; a narrow bottleneck fjord often clogged with ice; a dearth of land game; northeasterly gales; and the occasional *piteraq* wind which blows out every window in town. Angmagssalik's growth can be accounted for only by a geographical variant of Parkinson's Law, which states that the least likely place will always be the one picked by White Man to unload his burden. It was here, ten years after he failed to discover the missing Vikings, that Gustav Holm (now a captain) established the first trading post this side of Cape Farewell. It grew like a fort in Indian country; those who came to trade, or receive baptismal rites (Christ they accepted primarily because the missionaries offered them snuff, too), stayed on for the rest of their natural lives. Angmagssalik grew so much that in 1925 eighty-one Eastlanders decided to pick up their tents and move north to less crowded and more fertile hunting grounds at the mouth of Scoresby Sound, nearly five hundred miles from their closest neighbors.

Such a history suggests another Eskimo ghetto like Nuuq, yet Gustav Holm's erstwhile trading post is not ghettolike at all. A hodgepodge of jerry-built houses and jerry-built governmental structures (Angmagssalik is the administrative center of East Greenland), it has the erratic, leisurely sprawl of a frontier town or — given the abrupt hills rising up between houses — of the hulking, desolate, roche moutannée frontier itself. I've always liked Angmagssalik; it reminds me of towns as diverse as Fort Yukon, Alaska, and Everglades City, Florida. Towns where the works of man seem a bit confused to be part of the terrestrial estate and thus take on a quality of beguiling, ramshackle modesty. Also, it's hard not to like a place that provides the listener with such splendid outdoor concerts — the ceaseless music, day and night, of sled dogs.

Anything will get these dogs going: church bells, a bullhorn, a truck backfiring, spectators at a soccer match. But the hills of Angmagssalik truly come alive with the sound of their music when someone decides to empty his honey bucket in their dining area. Lucky dogs! Especially lucky dogs in the summer, when their masters try to keep their metabolism down by giving them only fish bones and the occasional fish head. The contents of the honey buckets provide not only an invigorating change of diet, but lots of important nutrients, too (whatever its defects, shit is vitamin-rich). And being bred originally from wolves, they celebrate their good fortune not by barking, which is sissy stuff, but by howling. They howl whole Wagnerian music dramas, choruses from the *Messiah,* Rodgers and Hammerstein hits, oldies but goodies, Beatles tunes, and the blues. One dog will begin a long riff and then the others will join in and after a while the whole town will be howling. The sound is intoxicating. On certain pellucid afternoons the visitor even feels like accompanying them with his own slightly less musical voice.

But I didn't accompany them. Maybe later. Right now I needed to ponder my options, never too many, now distressingly few, for a trip to Sermiligaq. There were, I learned, no flights in the foreseeable future, barring a medical emergency. Kong Oskar's Havn (where Angmagssalik lies) and Angmagssalik Fjord were so clogged with ice that the 200-ton provision ship *Einar Mikkelsen* could not make its usual circuit of the depots and outports in the district. A smaller ship, the *Kronprins Christian,* had tried to make the trip and now was hopelessly trapped in ice somewhere near Kungmiut. Sermiligaq, only seventy-five nautical miles from Angmagssalik, could have been in the next hemisphere for all its availability to the traveler.

I mentioned my determination to visit Sermiligaq to a Dane in the shipping office of the Royal Greenlandic Trading Post. "Why on earth do you want to go *there?*" he asked, claiming it was a backward if not completely recidivistic village. A bit dirty, too.

"That happens to be the sort of place I like," I told him.

"You are a social worker, perhaps?" He smiled.

After I pleaded innocent to this accusation, I had no choice but to tell him about my wanderings. Sermiligaq, I said, was a last place: far-flung, old-fashioned, and without amenities. Also, it was where Gustav Holm had thought the last of his Viking kinsmen were holed up. He didn't find any sign of them there, but this did not prevent him from naming two of Sermiligaq's offshore islands Eric the Red's Island and Leif's Island. Even if neither of these islands had witnessed a Viking visitation, I said, at least they had names that made me want to plant my feet on their remote, unadorned shores.

On a slip of paper the Dane wrote down the name of a man who had a boat substantially smaller than the *Einar Mikkelsen*. "Agility counts as much as size in an ice-mined sea," he said. "Talk to this man, and maybe he will be able to help you."

And so it was that I paid a call on a seal hunter named Avannaq Brodersen. Avannaq lived with his wife Else and assorted grandchildren (a daughter had died of leukemia) in a small corrugated iron house that hovered at a nearly 45-degree angle on top of its foundation of exposed bedrock. I walked right in because Eastlanders think it's quite impolite to bang one's fists on a person's door; you risk interrupting that person at some intimate domestic pastime, but that's not as bad as hammering on his house. (Also, the house might collapse.) As it turned out, I interrupted nothing more intimate than a casual snack: Avannaq and Else were seated at the table and eating pieces of seal nose in little square chunks by holding on to the whiskers as if they were toothpicks. This is the proper way to eat seal nose if you want to avoid greasy fingers.

Avannaq was fiftyish, compact, and walked like John Wayne, only not quite so slow. He had a broad predator's face, like a tiger's or a puma's, and was known around town as a *piniartorssuaq,* a mighty hunter, for his ability to find seals no matter how lean the pickings were for everybody else. He was one of the few men I'd met in Greenland who didn't drink — mostly, I think, because being a *piniartorssuaq* is a full-time job wherein your friends and neighbors praise your hunting prowess and then settle back with a case of Carlsberg while you put their praise to the test. A mighty hunter has to live up to his reputation or else he may end up losing the heroic sobriquet, which might drive him to drink himself.

Avannaq agreed to make the trip to Sermiligaq if I would stand in the bow and watch out for submerged ice. He was eager for the *kroner,* he said, because times were tough. The bottom had dropped out of the international sealskin market and a lot of hunters were using their extra skins for dog food. He mentioned a children's book called *Inuk's First Seal,* which the Home Rule government was mailing to the hundred thousand or so kids around the world who write letters to Santa Claus, c/o Greenland. The book was about a little boy who shoots his first seal, eats the meat, and then can't understand why his father is unable to sell the skin. Nor can his father understand why, since Greenland's seals are not at all endangered. Avannaq said he had even written a personal letter to Brigitte Bardot explaining the silliness of the sealskin boycott, but he'd never received the courtesy of an answer. He wondered whether she might not be illiterate.

All during my conversation with her husband, Else kept silent. Caucasian idlers like me aren't very common in Angmagssalik and she seemed to regard me in much the same way she'd regard a warty, green-skinned visitor from the planet Pluto. But just as I was getting ready to leave, she finally spoke. *"Ivikkatailiniakkutit muutammi,"* she told me. (Try not to fall out of the boat.) Fine advice! Then she scooped up the seal whiskers from the table and threw them out the window, along with some cigarette butts, to the dogs. On the way out I was lunged at by one of the dogs, who was perhaps weary of this cuisine and eager to sample a piece of Caucasian *tartare.*

I pitched my tent in Blomsterdalen, a lovely, surprising haven of grass with a good stream-fed salmon lake a mile or so west of town. A billowing snowy rise separated Blomsterdalen from Prestdalen, the next valley over, where the old *angakut* used to sit and gather their power. In Prestdalen they would rub stones together for days and days on end, until an enormous polar bear rose up from the cold depths of the lake Tassifiaq. So piqued would the bear be from the dreadful monotony of all that stone-rubbing that he'd eat the offender. The young novitiate *angakok* would be digested and then the bear would spit out his bones, which somehow would reacquire their flesh before hitting the ground. From this experience, repeated annually, the *angakok* got his training in matters of the spirit. After approximately twelve years, he'd be able to cut his throat and obliterate the gash merely by passing his hand over it; heal the sick with a gallimaufry of healthful herbs and glossolalia; and comb the hair of Nerrivik the Sea Goddess. The last man to get his training in this fashion was an *angakok* named Georg Uparsima, who lived in Angmagssalik until the mid-1970s. I had heard that Georg — a jolly old man despite having been spat out so often — possessed such a sense of the numinous that he could see the little souls that inhabited every pebble along Kong Oskar's Havn.

Prestdalen was a paradise of naked brooding mountains, spongy moss, and solitude. It was almost supernaturally quiet. I could hear a waterfall of last winter's melt-off ploshing down at least two miles away as if it were bathwater in the next room. I was so taken by all of this that I relocated my campsite there. I pitched my tent beside the polar bear's lake, where there were enough stones to equip an army of *angakut* — stones of every size and shape which, in their fecundity, reminded me that the East Greenland word for stone, *ujaraq*, is the same as the word for earth.

That night a blue diaphanous frost haze sat on the lake and made the snow-stippled mountains across the water look as if they were hovering on air. I began to feel a bit cold and since I didn't know any Greenlandic mantras, I attempted the ancient Tibetan art of *Thumo reskiang*, by which you can make your body so hot that flames shoot out the top of your head. Wandering Buddhist holy men have been known to survive subzero

Himalayan temperatures clad only in thin saffron robes by practicing *Thumo reskiang;* their meditative powers serve the same purpose as a whole down-lined wardrobe.

Thus I concentrated hard . . . harder . . . yet harder. With my outbreath I expelled pride, anger, and sloth. With my inbreath I took in big healthy drafts of the Holy Spirit. *Harder.* More pride and anger cast out; more Holy Spirit taken in. *Still harder.* But it didn't work. If anything, I felt a bit colder than I'd felt before I focused all my thoughts inward. But this was just as well, for if I had indeed become warm enough to abstract myself from physical reality, I wouldn't have appreciated the frost haze, of which I was now almost a literal part. Through this haze the golden glow of the Arctic night percolated so that the lake seemed full of waltzing, promenading, jerkily illumined ghosts. Right then and there I decided that if, in the absence of a good tasty *angakok,* the lake-dwelling bear wanted to feast off me, I wouldn't stand in his way. After all, there were only fifteen thousand polar bears in the world, and five billion of me. To let one of them devour my all-too-common flesh would, if only slightly, help adjust the grievous imbalance.

August 13. Late departure. A prevailing westerly has been blowing all morning, washing over the ice cap and giving Angmagssalik and Europe, respectively, their weather. But by early afternoon the wind has dropped, the sky is at once radiant and overcast, and Kong Oskar's Havn is like burnished silver. Avannaq says this kind of harmony between sea and sky, which he calls *ilimanarsertivaligajikkalivarimmit,* drives hunters crazy. The only crazy thing I can see about *this* hunter is the way he's packed the provisions, mingling our foodstuff lovingly with leaky fuel cans. Later, I know, I'll have to face up to this, but now I am more concerned with the state of our craft, a twenty-five-foot wooden dinghy dating back almost to Gustav Holm's day: torn canvas overhang, unpainted, nail-sprung, sheared-off exhaust pipe, and powered by a venerable single-cylinder engine. For some reason the word *Ulalujuq* (Whirlwind) is written across the bow . . .

We go past a skerry as wrinkled as an old work glove and twice as cheerless, except for a sky-blue carpet of squill along its shoreline, then pass from Kong Oskar's Havn to Angmagssalik Fjord, its panoply of icebergs and floes resting stock-still in the gray water. Swinging northeast, we sputter along beneath steep crosshatched cliffs, all creases and fissures, that put me in mind of the hide of a prehistoric lizard. Avannaq points to one of these cliffs and says it was once a girl who turned to stone — first her loins, then the rest of her body — because she wouldn't get married. That seems a bit severe, doesn't it? I say. *Tássaqa!* he replies: Every girl must marry. But before he can elaborate, two finback whales suddenly surface near the boat, their light-colored chevrons clearly visible, and both together send up fountains fifteen feet high from their blowholes. We're drenched with a spindrift that smells like rotting carrion in its terminal stages. Only smell I know that's worse than whale's blow is a whale's breath — a mephitic reek few whale watchers have ever smelled; if they did, they'd switch instanter to watching tiny odorless things like mayflies. Says Avannaq, after he's wiped the blow from his face: "That girl's name was Aleqasiaq, and she was my grandmother's eldest sister." Which makes him first cousin to a slab of monolithic rock!

Overnight stop at Qernertuarssuit, possibly one of Gunnbjorn's Skerries, so named because a Viking named Gunnbjorn Ulf-Krakason was blown off course in these parts around the year 900, inspiring Eric the Red himself to be blown off course — the square sails of Viking boats tended to encourage such deviations — to West Greenland ninety years later. Qernertuarssuit is like a mound of putty squeezed by a leviathan fist and then tossed to earth and allowed to harden. The magnitude and strength of that fist survive in the island's serried ridges and lifeline clefts. Avannaq's stepbrother Martin and Martin's wife Dorothe live here; only two other families and about sixty sled dogs are privileged to make a similar claim.

Martin, apparently, is another *piniartorssuaq*. Outside his red corrugated iron house are three *stativ* racks hung with long gnarled strips of seal meat. In his vestibule is a bucket of seal heads, their faces looking faintly surprised to be there. When we sit down to supper, we dine on boiled seal meat seasoned with somewhat rancid seal oil, with a side of seal blubber. Dorothe asks me whether I'd like a couple of sausages, made from seal intestines. *Qanorme!* I tell her, helping myself. No point in saving room for dessert; even for Dorothe, seal mousse or seal cheesecake would be an impossibility.

Just as the philosopher Ludwig Wittgenstein ate nothing but corn flakes at the end of his life (he thought if you found a food you liked, you should stick to it), so Martin and Dorothe apparently eat only seal, including the eyes lightly sautéed, which Martin believes is the only truly effective cure for a hangover. Yet unlike Wittgenstein, they have no choice in the matter, since the local trading post — two and a half shelves in a closet — stocks only sugar and ammunition.

August 14. Today the sea is green crème de menthe between pans and chunks of ice stirred around by the tides but never flushed out into open water. A few murres swim past us, their feet churning the sea like old stern-wheelers.

Going through the ice is like mapping out pieces of a seemingly endless jigsaw puzzle — except that each piece has ominous jaws. Constantly we zigzag around in search of open channels, "leads," which move like strong rivers. I'm also supposed to be searching for growlers just below the surface, a not easy task for my untrained eye. Around midmorning there is a loud resounding CRUNCH! Both of us are thrown off our feet and the boat itself is almost beached on a big blue floe lurking just beneath it like a gravel bar. *Tautûna,* says Avannaq, You did not see him. I did not, I reply, properly chastened. More ice; a football field suddenly cuts us off and obliges us to negotiate a big, looping parabola around it. You have this much ice in America? inquires Avannaq. He seems a little surprised when I tell him not so much. He'd always heard that America was the greatest land in the world in every respect.

Avannaq admits he hasn't done too much traveling. He was born in Kulusuk and at age ten moved to Angmagssalik with his mother (his father seems to be known but to God). Not once has he left this coast, a source of some pride with him, because many Eastlanders head west to Nuuq or Ilulissat for seasonal fishery work and return home brutalized, their slurred accents and backward manners mocked. He's also proud to have visited a doctor only twice in his life: once when he'd driven a nail into his thumb (the traditional remedy, soaking the thumb in hot coffee, hadn't worked), and another time when he had chilblains. Yet he was taken to an *angakok* at age four when he had whooping cough. The *angakok* recommended that he change his name, since coughs, like insomnia and rocks, were soulful entities with thought processes of their own and they tended to get confused if they found themselves installed in the wrong person. Avannaq's name was changed temporarily to Atsuiliq, Healthy Fellow, and the cough went away.

Ashore on Salissaq Island: a backbone of purple massifs and snowdrifts. At our approach a few chalk-white arctic hares hop off, not on all fours like ordinary rabbits, but on their enormous hind legs, like kangaroos — hind legs designed to keep them aloft in the loosest, deepest snow Sila the Weather Spirit can dump on their hopping grounds. Salissaq's dominant vegetation seems to be the bright vermilion lichen *Caloplaca elegans* (which Greenlanders call *sunain anaq,* sun shit) and the green

Carlsberg cans growing near the shore. In one cove I find a cache of birch and spruce logs, polished like marble after years of restless journeying from their Siberian homes along the Ob, Lena, and Yenisei rivers. In another cove my eyes follow a big white bird ploughing the water a few hundred yards away. Avannaq courteously informs me that it isn't a bird at all, no, not even remotely a bird, but a polar bear thoroughly unnerved by our presence and now swimming away.

Back on Angmagssalik Fjord. A flock of oldsquaw ducks shears across the high overcast, honking excitedly at the prospect of their winter vacation. The small black dots of seal heads bob up and down near a distant headland, but they're too far away, Avannaq says, to risk a shot. Besides, we're on their windward side, where they can hear us, rather than their leeward side, where they can't. Seals have terrific hearing but bad, rheumy eyes, which they favor by not sleeping more than a few minutes at a time, always bursting into wakefulness and glancing around, fearful of, primarily, bears and men. The best way to neutralize this fear, Avannaq says, is for the hunter to play seal himself. This he obligingly does for my benefit, barking, grunting, bobbing his head, and glancing nervously left and right, turning into one of the most convincing seals I've ever seen, except for the John Player's Special in his right flipper.

Kungmiut (pop. 100): a place of prefab houses serenely dozing its life away on an Archean promontory girded with mountains. This is the only village in Greenland, Avannaq observes, that doesn't have . . . *a single bathtub.* Oh well, I think to myself, I'll just have to go bathless for a while. During a period of polar exploration, Fridtjof Nansen once went three years without a bath and he later picked up the Nobel Peace Prize.

August 15. The sky's an eiderdown of fleecy wool. Another ragged skein of geese drives with collective purpose down the long slant of the season. We move along Ikarasaq, a narrow fjord scatter-patterned with ice like a madman's maze, full of cul-de-sacs and blind alleyways, from which we retreat, engine coughing and sputtering. At the present rate, we'll end up south at Cape Farewell, like this ice itself, rather than in Sermiligaq. Each time *Ulalujuq* has to stop and start up again, it kicks into an increasingly tubercular version of life. Could it be that the old crate's demise is imminent? Avannaq says not to worry. "If we get trapped in the ice, I have many bags of potato chips we can eat." Then he adds that this ice isn't really so bad; once he got stuck just north of here for some weeks and had to live off a ringed seal he'd shot. By the time he finally chugged back to Angmagssalik, his daughter had died, one of his sons had married, and Home Rule had been declared.

Awakened by a lone halo of sunlight, a ridge of knife-toothed mountains abruptly appears on the mainland, precipitous and doubtless unclimbable except by the gossamer motions of light rising along their slopes. We see a couple of ringed seals swimming among the floes with their characteristic sideways stroke, propelled by the force of their hind flippers. They're still too far away to taste Avannaq's .222 rifle. Being relatively blubberless this time of year, they'd sink before he could spear them with his staunchly outmoded iron-headed harpoon.

Fifteen or so miles from Sermiligaq we're beset by that most universal of human dilemmas: being stuck. Avannaq can't move *Ulalujuq* in any direction without running against jagged pans of ice. Just what I feared would happen. Just what he didn't fear. He withdraws into the canvas-covered hold and I hear the crash of crockery which indicates he's cooking up our lunch. No sense in letting a little predicament like being stranded here for the rest of our lives interfere with a good feed. Soon I'm slurping away at a whale stew in which I detect the slight garnish of engine fuel. Gladly would I slurp even if it had been marinated overnight in engine fuel. For the unalloyed air and low sky in these parts encourage the appetite;

subsistence rock and icy seas enlarge it; and a good bracing northeast wind at last liberates it from a preference for soy sauce over gasoline.

Several hours later the wind changes direction and the ice shifts slightly and we're able to inch ashore onto a gaunt, barren island called Qianarteq. We land near the ruins of a tiny settlement, now mostly a series of black sodden holes fringed with tent rings. A rich orange, nitrophilous lichen pours from several grave cairns as if it had been spurred to growth by the original bodies themselves. In one of these cairns I find a skeleton bent double in a fetal position; in another cairn, a skeleton perfect except for a missing left arm. In another I find two skeletons, probably mother and child, clasped together. How do you suppose they died? I inquire. No food, Avannaq replies. And then he cheerfully tears open a bag of potato chips.

August 16. Qianarteq. We're camped beside a litany of stones tucked together in a moraine like eggs in an egg carton. Earlier today Avannaq shot a two-year-old harp seal, called in Greenlandic *allatooq* to distinguish it from the young harp seal *(aataaveraq)*, the one-year-old *(aataatsiaq)*, the four-year-old *(aalattooruaq)*, the adult five-year-old *(aataarsuaq)*, and so on, each seal being classified with Linnaean precision according to its age. Later Avannaq defeats the flat taste of snow water by making Lipton's tea out of it. Read a Simenon mystery by flashlight. The ruckus of wind-blown rain and sleet against our tent tosses aside the possibility of sleep.

We occupied Qianarteq for three intermittently stormy days. On the fourth day the wind repositioned much of the ice, and we woke up to blue patches of open water. Quickly we gathered our gear and off we went — to Angmagssalik. For Avannaq needed to be back in this village by August 22 when he was scheduled to take a team of British glaciologists to the ice cap tongue near Isortoq. British scientists, he said, were very touchy about the clock. A year ago he was supposed to transport a group of geologists to Kungmiut, and when he was a day late they went off with somebody else.

I was so relieved that we hadn't suffered the same fate as our skeletal companions on Qianarteq that I didn't feel disappointed about not reaching Sermiligaq. I even looked on the bright side of it. The Vikings hadn't reached Sermiligaq, either. Thus our failure was rigorously in keeping with *their* itinerary. And as we bullied our path through more crazy paving of ice, Avannaq told me this story:

Fifteen years ago he had wintered over at Skjoldungen, a small hunting community established, like Scoresby Sound, to relieve the crowded conditions around Angmagssalik. Skjoldungen was set in the endless folds and crenellations of King Frederick VI's Coast a couple of hundred miles south of here. One day he was traveling with his dog team along a fjord two days out of Skjoldungen. He happened to camp at a small bight near the head of the fjord, and at his campsite he found another campsite, recently abandoned, which had the bootblack of char in the snow, a few tent rings, a midden of bones and discarded skins, and a rock that looked as if it had been shaped for flensing. He searched around, but found nothing of a more contemporary aspect, such as iron or spent ammunition. Other men had come back to Skjoldungen with similar accounts of abandoned campsites and one man said he'd even found a map carved in driftwood not unlike the old wooden maps Eastlanders used to carve in the days before Gustav Holm. All evidence seemed to point to a band of Stone Age hunters still living somewhere in these isolated fjords, a little nomadic scattering of people who had apparently never met any of their own latter-day countrymen. The survival of such a band was quite possible, Avannaq said, because almost no one had ever gone to that stretch of coast until Skjoldungen was founded in 1938.

It was quite possible, he repeated, eyes wide and full of wonder, and then he added:

"*Ela.* It would be best if no one went down there to discover them. They may be happy, they may be sad, but it would be best if they were left alone. Left alone to live or starve as they choose. *Attuniannaguk!* Best not to touch them at all . . ."

I promised never to touch them.

And then we were wheezing into Angmagssalik and my ears were greeted by a chorus of sled dogs howling a blues number they must have picked up from their human masters.

Chapter 12

CAIN'S LAND

Before they became captives of the ice, Eric the Red's successors sailed across Davis Strait to Baffin Island in search of timber for their boats, their roof beams, and their kitchen utensils. Baffin was, if possible, even less forested than Greenland, so they sailed southward into the Labrador Sea. The first Viking to visit Labrador itself was a man named Bjarni Herjulfsson. "A worthless land," Bjarni called it. But Thorfinn Karlsefni, who arrived in the year 1010, found generous forests and long sandy beaches shelving handsomely to the sea. Unfortunately there were *skraelings* (Indians) who took exception to Thorfinn and his men, particularly after being offered milk, a liquid they didn't possess the enzymes to digest. In fact, Thorvald Ericson, Leif's son, may have been killed by an Indian arrow in Hamilton Inlet; latter-day Labradorians, proud of their heritage, like to think so.

Legend has it that Custer's last dying words on the grass of the Little Big Horn were: "Well, at least we won't have to go back through South Dakota." I felt no more favorably disposed toward Nuuq, yet I had to fly back there from Kulusuk because Nuuq was the likeliest port in Greenland from which to hop a boat sailing southwest to Labrador. At the harbor I asked a French yachtsman if he could drop me off at any of several

Labrador outports en route to his native Normandy. *"Pas possible, Monsieur!"* he exclaimed. *"Pas possible!"* His response was so vehement that I wondered whether he had misunderstood me: I had said Labrador, not El Salvador, where some of Labrador's mail routinely ends up. I repeated my request, but he had indeed heard me correctly. Finally I figured he was just wary of the place described by his sixteenth-century countryman Jacques Cartier as "the land God gave to Cain."

Once again I had to violate my original plan of sticking only to slow boats. I flew from Nuuq to Frobisher Bay on Baffin Island; flew from Frobisher to Montreal; flew from Montreal to Gander, Newfoundland; caught the bus from Gander to Lewisporte; and took the Canadian National Maritime ferry *Sir Robert Bond* from Lewisporte five hundred miles to Goose Bay, Labrador. Every few hours the *Bond* lost her power and drifted precariously. And she seemed to stop and browse in every Newfoundland and Labrador outport before heading into Hamilton Inlet and docking at, as it is affectionately called, "Goose." Whereas the Vikings had made it to Labrador in no more than two days, it took me four and a half. Such are the vagaries of modern travel.

The first sound the visitor to Goose Bay hears upon his arrival is the ear-piercing BBRRRRRRRGGGGGGHHHHH! of warmongering aircraft overhead. I heard it even before I stepped off the *Bond.* I heard it again when I went into Captain Submarine, a sandwich shop also housing the editorial office for Labrador's only newspaper. I heard it yet again just as the clerk at the Corner Store rang up my six-pack of Molson's on his cash register.

Was Labrador at war with the rest of Canada? Was it sending off its air force to bomb Ottawa? Or was it merely at war with the province of Newfoundland, which indifferently administers it?

None of the above. These warmongering jets were part of

NATO's Operation Deep Strike, whose purpose is to strike targets deep inside the Soviet Union if and when the Free World decides to rearrange a bit of Soviet architecture. For nine months of the year Belgian F-16s, Italian fighter-bombers, and the F-4 Phantoms, Alpha Jets, and Tornedo CRIs of the West German Luftwaffe run low-level test flights out of Goose Bay into the interior wilds. (Interior Labrador is supposed to resemble the taigas and tundras of the USSR.)

BBRRRRGGGGGHHHHH!

I winced. The clerk didn't. Goose Bay, he said, was the odds-on favorite to become NATO's new Tactical Weapons Training Center, a fact that delighted him because the military always seemed to bring prosperity in its jet stream. I said prosperity was possibly the worst thing that could ever happen to a place.

"You ain't sidin' with those damned Injuns, are you?" he grinned.

"Gracious, no, friend. I was just thinking about a little town called Troy. You know Troy? It used to be prosperous, too. Now most of it sits in the Berlin Museum."

But I *was* siding with the damned Injuns. Or Innu, as they prefer to be called. They haven't been too appreciative of NATO's aircraft breaking the sound barrier (who'll put it back together again?) a hundred feet above their heads, hotdogging their camps with lean Pinocchio noses, and wreaking general environmental havoc on their traditional hunting grounds. A ragamuffin band of Northern Algonquin tribes, the Innu have wandered the Labrador Peninsula since shortly after the last Ice Age and are possibly the only Indians in the United States or Canada never to have signed any sort of treaty with White Man — neither a treaty exchanging their land for a mess of pottage nor one offering up their air space to state-of-the-art jets. Indeed, the Innu don't even regard themselves as Canadian citizens. Their mother country is a scraggly toupee on the bald surface of the Canadian Shield which they call Ntesinan (Our

Land). The mother of their race is not Her Majesty Queen Elizabeth but the muskrat; their father, that most predaceous member of the weasel family, the wolverine. But the muskrat population in Labrador has been reduced by low-level flying, and wolverines have been hunted almost to extinction. Sad to think of a race that will soon inhabit a biological orphanage.

Such people have a lot to get drunk about, so when I was joined outside the Corner Store by an Innu brave named Penote who looked like he wanted to get drunk, I obliged him with some of my Molson's. He accepted with a great toothless (sugar, flour, civilization) smile. Small beer, I knew, compared to the legendary spruce brew of home, which is a low-level flight in itself. The Innu boil the inner bark of spruce for an hour before mixing it with a concoction of yeast and molasses, which in turn they boil for four hours, adding raisins, prunes, vanilla extract, or whatever happens to be handy. I've heard of one man who added a bottle of Tylenol and another who threw in some athlete's foot powder. After this beer is imbibed, the fermentation process continues inside the drinker himself.

Penote and I were chatting in desultory fashion about White Man's duplicity when the noise of another overflight burst directly over our heads. BRRRGGGGGGHHHHH! Instinctively we ducked.

"Takutauat," Penote said, using the Innu word for Goose Bay, "is a lousy place."

I nodded in agreement. We decided to go where the only marauders were God-given black flies.

There is an Arab saying to the effect that if there are too many roads in your country, you die. If that is true, then Labradorians are among the best-situated people to live forever (the F-4 Phantoms notwithstanding), since there are only four roads in the whole peninsula. One of them leads west to the Churchill Falls hydro project and Smallwood Reservoir. A track of gravel and mud cuts through the iron-mine region of Wabush and connects with a railhead from Quebec in Esker. The third knits together a few southern fishing ports along the Strait of Belle Isle. Penote and I were setting off on the fourth, which leads along the shores of Lake Melville to North West River and the Innu settlement of Sheshashui. We walked at a slow, leisurely pace. It was a day warm as only summer days in the Canadian subarctic can be: a dense, humid, almost suffocating warmth that seems to suck up all the oxygen in the atmosphere and spit it back in the form of black flies. On the other side of Lake Melville, the Mealy Mountains were reduced by the heat haze to a smoky blue silhouette against an equally smoky blue sky.

Another jet roared overhead, leaving behind a thunder of bruised air.

"The exhaust from these planes," Penote said, gesturing skyward with his beer bottle, "it puts a green film on the water. The trout swim in it and you find them floating belly up the next day. The noise keeps the beavers in their lodges. It makes ducks and geese lay their eggs a month early. The mother caribou miscarries. The mother fox she's so scared, she eats her children. The porcupines have all committed suicide."

"Got a high-powered rifle? Why not drop a few of them out of the sky?" I recalled Farley Mowat's favorite childhood pastime of shooting at the American military aircraft flying over his native Newfoundland. He didn't down any, but maybe he was using the wrong type of gun.

Penote shook his head sadly. The Mounties had repossessed his rifle owing to an accident in which he had shot his brother-in-law. The real accident, I gathered, was that he hadn't killed the brother-in-law, but only plugged him amiably in the leg.

Seven miles out of Goose Bay a Hudson's Bay Company man picked us up and gave us a ride to North West River. I shared the back seat of his car with a trumpet-colored mutt that had the most ominous-looking muzzle I've ever seen.

North West River had some of the well-manicured charm of a New England village, with white frame houses set back among stately tamaracks and delphiniums blooming along the walkways. It had the first gardens worthy of the name that I'd seen in nearly three months; they were a bequeathal from Lord Strathcona (Donald Smith), founder of the Canadian Pacific Railway, who lived here in the 1880s and taught the locals how to wrest vegetables from their not necessarily generous soil.

Sheshashui, across from North West River's namesake river, had no gardens at all. A sort of halfway house for Innu not presently in the bush, it looked like Appalachia crossbred with a gypsy encampment and then struck by an earthquake. Canvas tents rested next door to uninsulated government tract houses, most of which seemed to have been trashed by indifference (I think I'd have trashed mine, too). The trash itself suggested that the Innu had been taking landscaping lessons from a crew of Greenlanders. Snowmachine treads, ragged fuel drums, cardboard, cast-off Evinrude outboards, bones, beer cans, flip-tops, and slops — all discarded with nary a qualm. Here and there piles of used Pampers showed the Innu were progressing with the years: formerly they used caribou moss for diapers. And I have never, ever seen such middens of Vienna sausage tins anywhere. They existed on a truly heroic scale. Oddly, I never once saw anyone actually eating a Vienna sausage during my stay in and around Sheshashui, which led me to conclude that the original tins must have been hardy perennials planted by that extraordinary gardener Lord Strathcona.

Penote said I could doss down with his family for the rest of my life just so long as they didn't have to fix my meals. He, his wife Annette, their three joyously dirty-faced kids, and a dribble-powered infant occupied a small prefab house. One whole wall of the house was overspread with a glossy technicolor poster of the pope — a tribute to the teachings of the

Oblate Fathers, who eased the Innu from a belief in scapula divination to a belief in Holy Water and transubstantiation.

I smelled not only the odor of unwashed bodies but also the odor of chamber pots. "Cozy place," I observed, "but when do you empty the honey buckets?"

"Only when they're full," Penote replied.

Instead of taking him up on his kind offer, I decided to pitch my tent on a little knoll two miles southwest of Sheshashui. It was a good, quiet spot, a suitable retreat after the jolt and jostle of my travels. I padded around barefoot on the caribou moss, an immensely abundant lichen as soft as a Persian rug and upon which footprints remain visible for days. I flushed up spruce grouse by the dozens; they flew to neighboring branches where they sat and clucked at me in astonishment. And I read *The Lure of the Labrador Wild,* Dillon Wallace's harrowing account of the ill-fated 1903 Hubbard Expedition from North West River to Ungava Bay. The expedition — comprising Leonidas Hubbard, Jr., a New York outdoorsman and sometime Sunday School teacher; Dillon Wallace, a lawyer; and George Elson, their half-breed Cree Indian guide — came to almost immediate grief when they mistook the Susan Brook for the Neskapi River, the standard Innu route into the interior. Their initial canoe crawl up the Susan took two tortuous weeks, traversed portages that were like mini-expeditions in themselves, and yielded a mere twenty-four miles. After leaving the Susan, they trekked around the barrens in search of Lake Michikimau, but only succeeded in getting themselves more and more hopelessly lost. Winter was approaching and now their wanderings became desperate. The land, not forgiving their original mistake, offered them only the starvation diet of boiled-down caribou hooves and antlers. Wallace and Elson at last got back to North West River, barely alive, but Hubbard died in the wintered-up barrens. His death was a monument to bad luck and a bravado nurtured by Kipling's poetry and the Bible.

"If Hubbard had a Labrador trapper with him," Gordon Maclean, himself an old Labrador trapper, once told me, "he would never have gone hungry. Nossir. No Labrador trapper ever starved or let his buddies starve. You ever eat porcupine? Now that's the sweetest meat there is. Ever eat lynx? Tastes just like turkey . . ."

Meanwhile Penote was doing a brisk trade in beaver glands. Even though every trapper swears by his own favorite bait (catnip, aftershave, fish oil, and so on), most of the Innu seemed to prefer Penote's special blend of two parts beaver glands to one part wintergreen and Old Spice. I was told martens liked this mixture so much that if they approached a trap and saw another animal there, they would scamper along the trapline until they located a free trap, eager for a death graced by a bouquet of beaver glands. Innu men must buy a new snowmachine each year due to wrack and ruin in the Labrador Wild, and a couple of high-grade marten pelts, marketed as sable, go a certain distance toward its purchase. The deceit seems like an appropriate revenge for the days when the Innu traded white fox pelts for priceless marvels like empty lard pails.

I had better things to do in Sheshashui than watch a man support his drinking habit by the sale of beaver glands. I needed to hire another native person to be my guide, this time for a trip to Porcupine Strand, a beach my Viking predecessors called Furdurstrandir (The Wonderful Beach). En route south they rowed and rowed past its glistening white sands; they must have thought the beach would go on forever. It does go on for nearly fifty miles, yet it has never been discovered by beach lovers, who would have to endure several rip-roaring days in a speedboat, followed by bushwhacking through woods and coastal tundra, to get there.

Porcupine Strand seemed not to have been discovered by the Innu, either. Their backcountry movements took them in the opposite direction, into Luftwaffe country, so very few of them had even heard of it. And a recent, rather unpleasant incident made me none too eager to knock on their doors or invade their tents. A young Innu man had taken a job wiring an officer's house in Goose Bay for cable TV, but his friends had ridiculed him so much about lackeying not just for a white but (evil of evils) a Military White that he quit the job. He came back and slashed up a tent belonging to a friend, along with that friend himself, who happened to be in the tent. All of a sudden I seemed to be a stand-in for the white officer and the whole thing became *my* fault. A man with a powerful beer breath and a cap that said *Instant Asshole . . . Just Add Alcohol* came up to me and, sticking a hard finger in my chest, asked me what I was doing in Sheshashui. "Researching the Vikings," I replied. Any other answer might have obliged him to reach for his knife, but this seemed to satisfy him and he walked away, belching contentedly. Then an hour or two later I was nearly hit in the head by a piece of the Canadian Shield, that mass of granite which is the world's largest exposed area of Precambrian rock. It had been flung at me by a little girl, maybe eight years old, who seemed to be acting at the behest of her elders. I looked at her and she smiled coyly. A few more years and she'd be a real catch for a man who fancied a concussion.

I began to despair of finding a guide. Then one evening Penote showed up at my tent with a man of about fifty-five, bandy-legged, broad of chest, with chamois skin and primal Asiatic features that suggested a journey across the Bering Strait 15,000 or so years ago. Despite a chill in the air, the man was wearing only jeans and a Montreal Expos T-shirt. He was introduced to me as George Apatet, guide *extraordinaire,* who took big game hunters into the bush in search of the black bear. George said he'd take me to Porcupine Strand if he could scout out certain burned river valleys, berry-rich and possibly bear-rich, along the way. Berries were ripe just now. Yes, indeed, very ripe. He pointed to Penote. As ripe as the nose on this drunken Indian's face.

I told George about my North Atlantic travels. At each mention of the places I'd visited, he clicked his tongue in pure wonderment. Having never set foot outside of Labrador, he told me his life's ambition was to see a cow.

September 4. An oddly put together day, like a bad print of a Bergman film, steely and dark, with a mist so fine it can be seen only when backlighted by stray sunbeams. George's twenty-two-foot motorboat nudges into inlets and navigates hummocky islands which, taken together, remind me of the Maine coast, all woodland frieze and rock scoured by retreating glaciers. Many of the trees along this shore are flexed back by the annual crack-up of winter ice: the ice jolts up onto land, throws its weight around, and intimidates the trees into a stance of horrified retreat. At one point we moor at a raggle-taggle pier and walk a mile inland to visit a trapper's old A-frame tilt, next to which a single swamp laurel stands, almost garishly red, advertising its charms for Labrador's occasional pollinators. There are no clamorous NATO planes in sight; no bears, either. But we do see a few piles of antique bear scat, white as talc.

Outside Waterton Lakes, Alberta, I once saw a grizzly walking upright and was struck by how closely it resembled a very shaggy human being. The resemblance increases, George says, once a bear is deprived of its shagginess. The skinned carcass of a bear is downright startling in its similarity to a person — the Innu used to see their own mirror image in that carcass and sit around smoking their pipes in its company (they'd put a pipeful of tobacco in its mouth, too). Unmarried women were not thought worthy to look upon this carcass, and only men were allowed to cut pieces of meat from it. The tail would not be cut off under any circumstance lest the bear's spirit be offended and seek revenge in the afterlife, which it shared with human beings. The Innu suspected that bears understood their language, just as some of them had once been able to speak fluent bear. But this fluency was lost now. George knows only one elder aged about a hundred who admits to a few grunts and snorts of bear lingo.

Back in Hamilton Inlet. More islands, bred in the glacial bone and lying low in the water like lazy whales with bits of shrubbery on their backs. Perhaps on one of these islands Thorvald Ericson ran afoul of George's ancestors. I suggest this to him, but the only Vikings he's ever heard of are the Minnesota Vikings, though he thinks Minnesota is in the state of Los Angeles. As we talk, he raises his 20-gauge and plucks a Canada goose out of the sky and then swings it, black-stocking neck like putty, from the water into the boat, an action that causes him to drop his cigarette onto the fuel tank. The possibility that this raunchy Gauloise will be my ticket to the afterlife crosses my mind, yet death is so casual in the North — enraged *qivigtut,* little girls with rocks, heading up the wrong river — that I might as well be frightened of isobars. (The traveler of a hundred years ago was warned: "Beware the awful isobar.")

We camp at a place called Baikie's Creek. The creek seems to have dried up and there's no sign of Baikie, either. Over supper George tells me this story from the days when all living things occupied the same realm of the spirit:

There was a hunter who hadn't eaten for quite a long time. He wandered the woods and barrens and at last came to a marsh where he saw a flock of geese in the water. "Brothers," he said, "let's have a big party tonight." "Okay, brother," the geese replied. That night they all gathered at a little clearing and the hunter told the geese to close their eyes — he wanted to play his drum. As he played, the geese danced around the fire. And as each bird passed him, the hunter grabbed it, twisted the neck, and killed it. Finally the leader of the geese caught on. "Brothers!" he yelled. "Our brother is killing us!" And now the few that were left opened their eyes and flew away.

"After that party," says George, taking a hearty bite of goose breast, "the Innu and geese were brothers no more."

September 5. Eskimo Island: a knuckly, uninhabited chunk of land in the eastern neck of Hamilton Inlet. Lots of driftwood lying about like the lashed leg bones of dinosaurs. The ground is covered with a heath shrub called Labrador tea, which makes a terrible tea, though George says the plant is eminently smokable and rolls a tea cigarette to prove it. Legend has it that Eskimo Island saw the last pitched battle — a rock-throwing, spear-thrusting, axe-hacking fight — between Labrador's Inuit and the Innu, time-honored adversaries. Yet a battle took place quite recently: delegates from the two groups (Inuit from Nain) were brought together for a caucus about low-level flying. They immediately went after each other's throats rather than after all those pernicious BRRRGGGGHHHs. In the same way, a starving red fox and a starving arctic fox will ignore a slab of ripe carrion and fight one another to the death in a triumph of ethnic prejudice over mere survival.

Warmish weather has brought back hungry multitudes of black flies. As with mosquitoes, only their women bite, an activity they perform with whorish abandon in order to get protein for their eggs. However, a case can be made for the black fly as a true conservationist, determined at all costs — often sacrificing its very life — to preserve the wilds, driving away the Great White-Assed Human Being so that Labrador will continue to be a last place. (Resource sniffers, I'm told, are the most susceptible of all, though they Abercrombie & Fitch themselves with veils, nets, and gloves.) For this, at least, the black fly ought to be declared provincial bird. George says you've just got to give in and accept them, histamine shock and all, as one of life's basic truths. Even now he seems no more concerned than if he were being politely drizzled upon. His face is a mosaic of black flies and yet he stands placidly by the Narrows between Eskimo Island and Henrietta Island, pulling one sea-run trout after another from the water. When I ask him what he plans to do with all those trout, he replies, "Eat 'em."

Dinner of sumptuous-flavored trout and potato buds flavored like sawdust. Soft, mazy evening. We talk about the Ungava grizzly, extinct in Labrador since (in all probability) 1914, when a final brown tousled pelt was brought to the Moravian Mission in Makkovik. George's grandfather may have been one of the last human beings to see an Ungava grizzly alive; he caught a glimpse of one gazing in sheer carnivorous pleasure at a herd of caribou around the turn of the century. But the introduction of firearms killed so many caribou that the grizzly lost its main source of food and ended up collectively in the afterlife.

Idea: Set aside a few thousand square miles of Labrador for this bear in order to placate his doubtless morose Spirit. Kick out all the mankind creatures who have not yet been driven away by the black flies or eaten alive by the Labrador bulldog, a species of horsefly less prevalent than his black brother but capable of far more individual flesh-tearing violence. And hope against hope that this munificent gift will appeal to the old extinct fellow and he'll clothe himself with fur and stalk the barren grounds once again as in the more fruitful days of yore.

September 6. Head winds keep us at Eskimo Island for another day. Trout for breakfast. I listen to the bravura orchestra of the wind and read some of E. H. Carr's biography of the

Russian anarchist Bakunin. Bakunin, it appears, had the regrettable habit of posting ciphered letters with the code enclosed in the same envelope. Also, he forced himself to eat bad food because Russian peasants ate bad food. I ask George what was the worst food he's ever eaten. He answers without hesitation: chips cut from his own snowshoes. Didn't the Innu used to eat caribou droppings in the old days? Oh, caribou droppings, he says, they're not bad at all if properly cooked . . .

We passed through the nautical jaws of the Narrows and came out in Groswater Bay, then bounced along the jigsaw Atlantic coastline to West Bay, a bygone fishing community now being reclaimed, for the most part, by coastal tundra. Along with hundreds of other teardrop-sized outports, it fell to the hatchet of Newfoundland-Labrador premier Joey Smallwood's Fisheries Household Resettlement Program in the 1950s and 1960s, a program not unlike the G60 Policy in West Greenland. Backed by joint federal/provincial funding, Smallwood took a militant carrot-and-stick approach to the issue of rural resettlement. He shelved essential services and at the same time offered outporters incentives and jobs if they'd pick up their lives and move to designated "growth centers." Smallwood, once an unsuccessful pig farmer, thought he understood fishermen; he thought they'd prefer factory work and hygienic conveyor belts to their smelly gear. He was mistaken. Like the human cargo in Greenland, the refugees cordially detested their new surroundings and used every seagoing convenience at their disposal to revisit, if only for the summer, their native bedrock.

West Bay and Fish Cove offered the only safe moorings for Porcupine Strand in this stretch of coast. George chose West Bay because he wanted to drop in on two old friends, Stanley and Helen Mugford, victims of the Smallwood purge. Official residents of the downcoast metropolis of Cartwright (pop.

685), they returned each summer for the inshore fishing and stayed in the same frame house, once white, now gray, that had been their year-round abode in the old days.

We tied up the boat and climbed a rocky embankment to the house, where the Mugfords received us. Both seemed to be composed of equal parts wry humor and traditional fisher-person's bile.

"We left, my b'y," Stanley told me, "but we shift back here to prime our heart pumps." He was an angular gray-haired man who smoked hand-rolled cigarettes and told me he'd read so much about the dangers of smoking that he'd had to give up reading altogether. He said he hated Goose Bay, he hated Cartwright, and he didn't even like West Bay so much any-more — too many long-liners.

"Them long-liners," Helen butted in, "they wait and see where fellas like Stanley does their jiggin' and then they swarm all over the grounds. They got nets three miles long, the sonsabitches."

A dusky double-chinned mountain of a woman, she reminded me of an Irish tinker colleen I once knew who got into a fist fight with a Civic Guard and sent the poor strapping fellow to the hospital with much bodily damage. Helen's background, she proudly informed me, was Newfie, Eskimo, French-Canadian, and Scotch, all of which seemed to show up in one part of her or another.

The Mugfords offered us brown bread and roasted periwinkles, which they called "wrinkles." "By God," Helen roared, "they's no candy nicer than wrinkles!" We sat on stools around an old tin box stove whose door was fastened on a hinge of wire with small interstitial loops and a handle that was little more than a larger loop. Once upon a time trappers used to equip their log cabin tilts with this sort of stove. It warmed quickly and threw off considerable heat, though it lost its heat just as quickly; Helen was always scrambling to her feet and tossing in birch logs and dry spruce to feed its seemingly insatiable appetite. On one of these trips she stopped and switched on a shortwave that had plainly seen more elegant days and now could not even broadcast, only receive. We were treated to a dialogue between two truckers caught in a traffic jam outside of New York City ("Fuck, Blue Devil, man. Like, I've been sittin at this fuckin' underpass for two fuckin' hours. Over and fuckin' out, man"). The Mugfords, eighteen miles from their nearest neighbors, never seemed to tire of such casual involvements with their fellow man. Only last week, they said, they picked up a man in Michigan who declared he had beamed into the radio control of a flying saucer.

"Porcupine, eh?" said Stanley, rolling himself a cigarette from tobacco warmed on the stovepipe. "I heard the Vikings was there onct, but they've scarcened up now. Plenty of bruins around there, though. Seen bruin scat all over the place. Heaps and heaps of it, my b'y. By God but they's ever gassy creatures, bruins is. You ever hear one of them fart? *Holy Moses!*"

Said Helen: "Knowed a man who taught a bear to drink from a bottle of rum. That'd be Uncle Bert Saunders. Remember Uncle Bert, Stanley? He froze to death."

"I remember him, girl," said Stanley, rolling himself another cigarette.

"As long as I live, I'll never forget that bear drinkin' rum."

Tough, generous, crusty, rudimental people. I left Stanley the Bakunin book on the chance it might give him a few anarchistic ideas for weathering the urban winter in Cartwright. In return he wanted to give me an old jump-spark engine I'd admired, but I told him it wouldn't fit into my pack, so he gave me a bag of wrinkles instead.

Now George and I resumed our trip to the bear-scat-littered sweep of Porcupine Strand, following a path behind the Mugfords' cabin that immediately dwindled into nothingness and then reappeared as a series of lemming trails in the moss. As we walked on, we heard the noisy hawklike whistles of a pair of whiskey jays and then saw them squabbling over a dead lemming, which one bird would steal from the other and the first steal right back again. Soon we came to a bleached ancient boneyard of a forest destroyed by fire — fingertips of splinters aiming skyward, eerie elongated stump faces, charred brash strewn all over the ground. George had been complaining about six miserable weeks he'd once spent as a salmon gillie at a fishing camp on Eagle River. Never again, he said. But he fell silent — we both fell silent — at this implicit reminder that men and trees share the same mortal roots. The only sound I could discern anywhere was a woodpecker trip-hammering a hole in one of the dead stumps.

After a few more miles we arrived at a marshy plain with hundreds of pitcher plants, their bristles pointing into their hungry gullets. I watched a bee slowly crawl to oblivion. Farther on we came to a patch of bakeapples *(Rubus chamaemorus)* and I stooped down to achieve a more pleasant sort of oblivion with these reddish orange berries. Their drupelets are said to have a baked apple flavor, but the hint of sumptuous decay even when not quite ripe reminds me more of gusty, fermented, red-blooded cherries.

Meanwhile George had located a trail of bear tracks beaten firmly to the bakeapple patch along a high, dry ridge. Bears are lazy travelers and will follow well-trod routes (another human similarity?) rather than blaze new trails on their own. Along this trail were numerous piles of bear shit, full of berry seeds and looking for all the world like black caviar and red caviar blended together. George seemed quite pleased, although by now we were at a place far too remote for his big game hunters. He bent down and picked up a drying clump of scat and scrutinized it carefully, saying, "This bear's been eating ants, too." While he stood there with that turd in the palm of his hand, he seemed possessed by a strange sort of harmony, as if he were suddenly renewing his kinship with the bear kingdom. Then he placed the turd gently on the ground.

We tramped across a muskeg of sphagnum moss and sedge rolling eastward toward the sea and then at last we came out on Porcupine Strand — a remarkable sight! It stretched seemingly without end all the way down to the spoon-shaped headland of Cape Porcupine, mile upon mile of bygone rock crushed to sand by coastal ice and then washed ashore, perhaps the longest, emptiest, most breathtakingly tourist-free beach in the whole world. Grains of yellow sand flashed and sparkled even though the sky was pitch gray. After the neutral colors of muskeg, the sparkle made my retinas ache. But after sinking into so much moss, it was a distinct pleasure to sink into sand.

As George and I walked down the wonderful beach, I noticed more piles of bear shit and several dead beluga whales, which doubtless were a culinary attraction for the bears. I also noticed spars and planking from long-outmoded sailing vessels, along with the usual aluminum cans, detergent flasks, rubber boots, rubber gloves, shreds of netting, and tangles of plastic and monofilament. To my consternation I then began to notice gaping craters in the sand. They were obviously not of Viking origin — unless the Vikings possessed aircraft capable of carrying heavy bombs. Soon I was finding pieces of sheet metal, bomb casings, and rusted-out bits of shrapnel, some resembling spare auto parts, others looking more like convulsed human bodies. One torpedo-shaped object looked especially fearsome and possibly even live (I wasn't curious enough to find out).

Where all this military detritus came from, I had no notion. Maybe it came from Cain. It was bad enough that he had killed his brother, but when God gave him a sort of suspended sentence by exiling him to Labrador, he had returned the favor by trashing it. He had turned the Wonderful Beach into an unwonderful battlefield such as you might expect to find in Vietnam, Lebanon, Bull Run, the Somme, or maybe even Narsarssuaq, but not in a place as removed from normal human hostilities as this. Not *here*.

George had observed dummy smoke bombs dropped all over the bush northwest of Sheshashui, so he was rather unperturbed by all this rusty hardware. He bent down and picked up — wonder of wonders! — a tampon applicator. I had been paying so much attention to all the other junk that I hadn't noticed the beach was littered with them, though not quite so abundantly as Revere Beach (where they seem to breed), just north of Boston. There was no telling where these had originated — Goose Bay? Montreal? Qaqortoq, Greenland? — or how long they had bobbed around in the high seas searching for a suitable landfall; for nowadays the human species flushes so much refuse into the water that ocean currents are no more useful than astrological charts for predicting where something came from and where it will end up.

"What do you call these things?" George said.

"Beach sticks. Tampon applicators."

"We call them poor man's birchbark. They make a real fine fire. In the rain, plastic burns a whole lot better than wood." And then he proceeded to gather up a batch of them, together with other plastic discards such as Josol fuel jugs and liquid detergent flasks; mixing them all with weathered pieces of driftwood, he soon had a roaring if slightly pungent fire going. We sat around this fire as if it were a snug little camp in the woods, and George began telling me about his son's muskrat harvest, which last year dropped from seventy-five to twenty, because the NATO jets were causing the muskrats to commit suicide.

"Suicide?"

"Yes. They just turn on their backs and die."

In an hour or so it started to rain, a cold persistent drizzle that rolled in off the gray Labrador Sea and cut short our day at the beach.

* * *

Back in Goose Bay, I asked Colonel Jonathan David at the air base if he knew anything about the origin of the bombs on Porcupine Strand. Being an Ottawa man, he didn't know the whereabouts of Porcupine Strand, but he suggested the bombs might have been dropped by the Americans during the old Strategic Air Command days. The coast, he said, was off limits to NATO planes. Then I went to the harbormaster, who thought they might be vintage World War II mines long since washed ashore. But he couldn't tell me for sure or explain the presence of the torpedolike object — obviously not a mine — that I'd seen stuck in with the other rusty ordnance. No one could tell me for sure. After a while I changed my mind about Cain and decided all that hardware simply grew there, like the Vienna sausage tins: a new metallic species of plant as befits a time of war.

Chapter 13

LAST PLACE

SEPTEMBER 15. On the *Sir Robert Bond* again. A murky day. We follow the same route I took with George: through the Narrows, into Groswater Bay, and down the coast. Once we reach the open sea, the strong south-setting effect of the Labrador Current, which deflects the Gulf Stream to Iceland, moves us along briskly. A pod of pilot whales follows us, gathering on the starboard, diving under, surfacing on the port, diving under again, and reconverging on the starboard. The elders surface with a stately *whoosh*; the young, like a cork popped from a champagne bottle. One of the crew tells me that Newfoundlanders call these whales potheads, from the shape of their head, though it's obvious, he says, that these particular potheads are stoned.

We dock briefly at Cartwright, where a family comes aboard, all of whom — father, mother, son, two daughters — look like Abraham Lincoln. Another new passenger, a man from Toronto, tells me he's given up sex because he's afraid he'll get cancer from all that friction (his outlet, he says, is tae kwon do). As the boat pulls away, I notice a funeral procession winding out of a cemetery just as a similar procession of children is winding out of a drab-looking concrete schoolhouse.

Muddy Bay: where food tycoon Clarence Birdseye had a fox and mink farm and learned the secrets of quick freezing (first fish, then cabbage) from a local outporter named Garland Lethbridge. Later Birdseye made a fortune from these secrets. Yet it is to Garland Lethbridge that the devotee of frozen broccoli, frozen egg rolls, and frozen TV dinners owes his allegiance.

Chilling fog. More potheads. Icebergs grounded on the coast. Icebergs not grounded on the coast. Icebergs like grandiose Himalayan peaks. Icebergs like little Irish drumlins. Icebergs like the foothills of the Rockies. Not surprisingly this shipping lane is known as Iceberg Alley, and a previous coastal ferry, the MV *William Carson,* was sunk after being stove-in by a hulking alley dweller.

Departed the *Bond* at St. Anthony (pop. 2,900), "capital" of the Great Northern Peninsula of Newfoundland and Lambarene of the North. It was here, in the early years of the century, that Sir Wilfred Grenfell located his first mission. Like Schweitzer, Grenfell healed the sick, promoted Christ, and banned alcohol. Not just alcohol; he also tried to ban spitting, a prophylactic measure he thought might reduce the high local incidence of tuberculosis. Lucrative bonuses were paid to rug hookers whose rugs read "Don't Spit" instead of the customary "God Bless Our Home." Today St. Anthony is home to the Viking Mall, Viking Video, Viking Laundromat, and the Vinland Motel. It's also the end of the Viking Trail (formerly Leif Ericson Highway), specifically designated by the province to honor the party of seafaring visitors from Greenland who long ago docked on these shores.

The Great Northern Peninsula sticks out from the rest of Newfoundland like a long accusatory finger pointing north at Greenland. The Labrador Current joins land winds to make it a shooting gallery of bad weather; westerly winds off the Gulf of St. Lawrence lend their teeth to its already gnawed surfaces; and the Long Range Mountains isolate it from the rest of the province. The first road, called the Devil's Highway, appeared in the 1960s and was immediately pronounced worse than the previous lack of road. As recently as the middle years of the last century, the peninsula did not have any permanent settlements, only seasonal stations built by French fishermen from St. Brieuc, St. Malo, and St. Jean-de-Luz. Then the French began to hire domestic *gardiens* to look after their gear in the winter; thus the first threadbare communities in the peninsula came into existence. These *gardiens* tended to be of the same stripe as the men who had settled the rest of Newfoundland: deserters from the British Royal Navy, luckless fishermen from Devon and Cornwall, even more luckless victims of the Irish potato famine, and fugitives from justice. Just the sort of people who could blend their hard-bitten genes with this hard-bitten terrain and come up with something like a union. Their descendants persevere on the peninsula to this day, and not a few of them remain, after a fashion, fearlessly marginal, running their cockleshell dories in the teeth of twentieth-century winds.

It was to the little outport of L'Anse aux Meadows (an English corruption of L'Anse aux Meduse, Jellyfish Creek), at the very northern tip of the peninsula, that the Norwegian explorer Helge Ingstad came in 1960, having already foraged much of the eastern seaboard from Labrador to New England in search of Leif Ericson's Vinland. As described in the Sagas, Vinland was a place chiefly remarkable for its wild grapes and/or berries, depending on how you read the word "vin." In the past any wag who found odd rock inscriptions in his back yard seemed both willing and able to claim Vinland for himself. But Ingstad was going about his search more rigorously if not more personally (he had an ancestral interest in the Vikings), and his mind was finely tuned to northern landscapes from years in Alaska and East Greenland. In L'Anse aux Meadows, he found the covering of grasses the Vikings would have needed for their sheep and cattle. He found islands, streams, and a lake that came close to matching Saga descriptions of Vinland. And he found a farmer-fisherman named George Decker, who regaled him in his rich Newfie brogue with stories of men and women in strange old-fashioned costumes who would suddenly appear out of nowhere and help local people with their haying.

Someone who can tell a good ghost story should never be ignored. Helge Ingstad attended to George Decker, who led him to a group of soddy humps ("Indian camps," Decker called them) situated on a broad ellipse overlooking Wreck Bay. They seemed more promising than other soddy humps Ingstad had been looking at along this same coast. And indeed they were. Between 1961 and 1969 Helge and his archaeologist wife Anne-Stine excavated a site on George Decker's pastureland that yielded eight sod houses, an old cloak pin, a bone needle, some fossilized bread, and the flywheel of a spindle similar to flywheels still being used in Norway. Later digs showed the presence of Maritime Archaic and Beothuk Indians before, after, and maybe during the Viking tenancy. Sixteenth-century Basque whalers may even have camped on George Decker's land, which was clearly a place of ancient resonance, full of multinational ghosts. In 1978 UNESCO declared Decker's land and its environs a World Heritage Site, a decision that made at least a few of the hard-bitten citizens of L'Anse aux Meadows reluctant to drop coins in a UNESCO cup ever again. As one man put it: "Our fathers had always hunted and trapped on that land. If we can't hunt and trap there ourselves, then bugger the Vikings . . ." No doubt the Vikings would have agreed with him.

In St. Anthony I rented a car (an act I can't defend except on the grounds that it was beat up and I was feeling a little beat up myself by this time) and drove thirty-five miles north to L'Anse aux Meadows. I passed through outports like Gricquet, St. Lunaire, Gunner's Cove, and Noddy Bay — villages of doughty clapboard houses and gray fishing stages held in a breath of stillness over the edges of small coves and the incessant sea. In Gricquet I saw a pretty young girl in a bright red dress gaily swabbing down the family truck; near St. Lunaire I stopped for lunch at a restaurant that advertised "Three Coarse Meals."

Being a connoisseur of bad roads, I wasn't gratified to see that the classic washboard ruts and tooth-loosening gravel of Newfoundland Route 81 were getting paved over for the very first time. There would be no more stories of cars with front *and* rear windows shattered by the same high-flying rock; no more stories of rocks penetrating steel floorboards; no more stories of mountainy corrugations followed immediately by bottomless potholes referred to by locals as "yes ma'ams." At long last these little outports would be hooked up with the decline of the West. Now they would suffer a different sort of resettlement from the type invented by Joey Smallwood: from Spillar's Cove to the Viking Mall in one quick blur (good roads make good malls). Small comfort that the class B gravel used by the road graders was still capable of playing old-fashioned hardball with an automobile's transmission.

The blacktop had not yet reached L'Anse aux Meadows, which was located at the very end of the gravel and was a higgledy-piggledy of small white houses backed by muskeg and peering out bravely on the Labrador Sea. Nobody would ever have had the ingenuity to live here were it not for the blockade of Great Sacred Island and Little Sacred Island safeguarding Medée Bay from the angry improvisations of North Atlantic storms. And until 1870, when George Decker's grandfather arrived, nobody did live here, except that motley of ghosts.

Down by one of the stages several fishermen still in their oilskins and thigh-high boots were waiting for the buyer who came each day from Gricquet. One of them had a beaky nose like a puffin's; another's hands had been ripped up by a lifetime of hooks and sizzling ropes; a third looked a bit like a surgeon deteriorated into a desert rat. They were engaged in one of the most popular pastimes in Newfoundland, swapping Newfie jokes:

"A social worker came 'round here and knocked on one of the doors. A young girl answers. 'Could I talk with your father?' he says. 'Fat'er 'e ain't home, 'e's in jail,' says the girl. 'What about your mother, then?' 'Well, mot'er's in the lunatic.' 'Do you have any brothers?' 'Yessir, der's t'ree. Tom, Dick, an' 'arry.' 'Can I speak with them?' 'Well, Tom took after fat'er, 'e's in jail.' 'How about Dick, then?' 'Well, Dick took after mot'er, 'e's in the lunatic.' 'And Harry?' 'You can't speak wit' 'arry, 'e do be at Harvard.' 'Harvard! What's he studying?' 'Oh no. 'e ain't studyin' — dey's studyin' 'e . . .' "

They laughed. I laughed. The kittiwakes circling the harbor laughed. Even the dead dogfish lying on the pier seemed to laugh. For backward people (Newfies, Polacks, Kerrymen, and Hafnarfjördur-ites) are prime suspects — even to themselves — in an age when national self-esteem tends to be measured by technostunts in space. No Newfoundlander has yet traveled to the Moon except the healthy way, through a surfeit of booze. In outports like L'Anse aux Meadows, Newfoundlanders still fish inshore with traps and handlines, two to a dory, as their fathers did and their grandfathers and even their forefathers in some watery corner of preindustrial Britain. The penury of the sea they prefer to the penury of a Toronto sweatshop. Another Newfie joke.

I fell in with the desert rat. His name was Billy Heddison, and the language he spoke was an almighty grab bag of Devon English, Irish English, lockjaw, and the sea; his conversation was peppered with hard Old Country words like "yaffle" (a lot of fish), "curry" (fish offal), "duckish" (dusk), "binickt" (ill-tempered), "ballyrag" (abuse), "flea-lugged" (hare-brained), and "scrunker" (stranger), which seem to have survived, like the Norn words on Foula, through sheer visceral force. As I heard him complain about the lack of market for any fish but cod, I could understand why the *Dictionary of Newfoundland English* (put together, it is said, so that Newfoundlanders can communicate with each other) runs to 650 pages. I also remembered talking to an old Newfoundlander in Goose Bay with a pain in his "pole." He said his sweet young nurse from Hamilton, Ontario, had blushed at this — she had no idea he meant his neck.

" 'e goahng t' de Vikin' site, my b'y?" Billy asked me. Then he pointed a thumb shaped like a furze root at himself and informed me that he was a Viking, too.

In a way, he was. His grandfather was one of several Scandinavians — Norwegians and Lapps — whom the missionary Dr. Grenfell brought over in 1908 to herd his reindeer. Grenfell brought the reindeer from across the Atlantic too. The little English doctor (beneath his polar garb he always wore Oxford University underwear) was determined to replace the Eskimo sled dogs of Newfoundland and Labrador with a more cost-effective animal. Reindeer could be trained to harness more quickly than dogs; the fur made a fine cold-weather sleeping bag; a person's transport, if necessary, could be eaten and the meat tasted better than haunch of dog; the doe's rich milk fought tuberculosis; the dried tendons could be used for sewing; and reindeer could graze forever on moss and lichen rather than depend on human handouts. But Grenfell's imported reindeer were hopeless. Either they would dash off with their drivers in pursuit of an unusually succulent patch of moss or they would dash off to inconvenient hillsides where they thought the rest of the herd might be congregating. Then the Lapps got homesick and went back to Lapland, though not before they carved some very rude remarks about Grenfell (remarks thought until a few years ago to be the sacred script of medieval Irish monks) in the rocks of Gricquet. Max Heddison stayed on, however, and settled in Gricquet with a Barnett woman who bore him ten children. One of these children, Billy's father, was posed in a famous photograph by Grenfell inside a seedy blimplike sweater with a little Colbourne girl from L'Anse aux Meadows — to show prospective mission patrons just how poor were the children of the Great Northern Peninsula of Newfoundland. And when he grew up, Billy's father married the Colbourne girl who had shared the seedy sweater with him.

After the fish buyer came, I went back to Billy's for supper. His house was a block-style bungalow that he had knocked down and rebuilt three times, each time picking out the nails, straightening them, and fitting the head all over again, which he said made them better than brand-new store-bought nails.

Nails have always been one of my favorite human inventions: so simple, yet the complexities of most edifices would collapse without them. While I was admiring Billy's born-again nails, a large German shepherd (you never see Newfoundland dogs in Newfoundland — people think they're stupid) came exuberantly to life, rushing at me and then snapping at my heels and hindmost parts. I always seem to have this effect on dogs. Perhaps they know that if given half the chance, I'd cut the cord that ties them to their cushy domestic lives and set them loose in the wilds where they could do a decent day's work of predation on behalf of the species. No matter. This particular animal changed his tune once he entered the house and soon was attempting to dry-hump my left leg with a determined rhythm.

The family was already seated at the dinner table — four brawny sons and a brawny teenage daughter. It happened that Billy's wife Nance was a cousin of a trapper I'd met in North West River, and they had the same expansive forehead and gooseberry eyes. Now I sat down to a plate of the national food of the Newfoundland outporter, fish and brewis, which consists of salt cod, hardtack softened in water, and crisp-fried scrunchins of pork all scrambled together and simmered on the hob.

"Buyer came late again today, eh," said Nance. "Only a bloody quintal of fish anyways," Billy replied. "Saw a moose in the schoolyard," the youngest son said. "Shot me a half dozen bunnies," said another son, Eskimo dark. "Shot me a brace of tickleaces," the eldest son said. "Bloody draggers with their bloody otter trawls bloody near've cleaned out the sea," said Billy. "Pass the milk," said the daughter. "Pass the seal," said the third son, and from a big platter took a side of baby harp seal, complete with crystal bones, so tender they were as edible as the meat itself. "You with the roads?" the Eskimo-dark son asked me. "This man's a writer from the Boston States," Billy told him. "Only book worth readin' is *The Lure of the Labrador Wild,*" declared the eldest son. "Ought to teach that book in the schools," said the Eskimo-dark son. "Instead of algebra and geometry," added the third son. "Pass the milk," said Nance. "Pass the seal, my son," the youngest son said to me. "They's a man in Snappy's," observed the eldest son, "an' he sells seal cockadoos to the Japs." "Eat anything, them Japs," said Billy. "They's for sexual purposes," the eldest son said. "Them Japs is real hard up, eh," said Nance. "They grind 'em down first," said the eldest son. "An' them bloody draggers," said Billy, "they's getting a better price for their fish than the little man." "Think I'll go out and shoot me some tickleaces," the Eskimo-dark son said. "Think I'll shoot me a bunny or two," the eldest son said. "Pass the fish, my son," the third son said to me. "Pass the milk," said the daughter. "Why *are* the draggers getting a better price?" I asked Billy. "Because they keep the fish plants in business, that's why. The little man with his little catch, he don't matter two bloody knobs to the fish plants." Said the youngest son: "You hear about the Newfie that took his nose apart to see what made it run?"

In this part of the peninsula only a few outports such as Fortune, Lock's Harbor, and Ireland Bight were made one with Nineveh by Joey Smallwood. Briefly in the 1960s there was talk of transferring the one hundred inhabitants of L'Anse aux Meadows to Noddy Bay, five miles away — of course, they hated Noddy Bay, bloody awful place, an hour farther from the good fishing grounds. Luckily nothing ever came of it. I got the distinct impression that if Smallwood *had* tried to remove Billy and his family from their thrice-rebuilt home, he'd have ended up as fish bait (Billy, on Smallwood: "The best part of that man dripped down his mother's leg"). The Newfoundland cod is a voracious creature that will devour almost anything dropped into the sea. Scissors, knives, keys, watches, oil cans, beer bottles, sea boots, oarlocks, entire lobsters, even Holy Bibles — all have been found in their egalitarian stomachs. It would not be too far-fetched to imagine a cod striking at a scrap of overweening politician.

Next morning I walked along the rocky shore to the Viking site. At one point I almost tripped over a dead sheep, belly torn open and crawling, not with maggots but with dozens of tiny frenetic red-stippled crabs. Dark as pitch, the cliffs of Labrador loomed across the Strait of Belle Isle. Great Sacred Island loomed only a few miles away, with the rusted-out wreck of the 5,000-ton cargo vessel SS *Langlecrag* draped along half its shoreward boulders, providing the main hardware supply depot for the fishermen of L'Anse aux Meadows looking for metal to bend into anchors.

Soon I was walking among the boardwalks, wooden bridges, and neat wickerwork fences of the site excavated by Helge and Anne-Stine Ingstad. Immaculate bronze plaques awarded titles to each grassy depression along Black Duck Brook: furnace and smithy, bath-house, charcoal kiln, workshop, dwelling, and outbuilding. The muskeg was as closely cropped as a suburban lawn, and the reconstructed Viking longhouse looked at least as comfortable as a suburban longhouse. Somehow this fine job of manicuring and renovation by Parks Canada seemed a little alien, even exotic, for the curious-spirited adventurers from Greenland. All that symmetry seemed wrong for their cold aquatic minds. I think it would have helped if the site were a little messier; messiness would have suggested more of what it's like to brave the frigid Labrador Sea in an open boat and then chance upon a block of land not previously on anyone's map.

But I loved the berries that carpeted much of the site. Cranberries, partridge berries, crowberries, and tundra bilberries. As they are now, so they would have been around the year 1000, and a Viking wanderer would have been struck by their profusion. That same wanderer could have bottled quite a supply of Château Vinlandia for his wine cellar and spent many a chillsome wintry night traveling to the Moon without ever leaving his soddy home. Whether these berries identify the Ingstads' site as Vinland is another matter. It was at least a winter camp and possibly a base for exploring regions farther to the south, but the Sagas are typically vague about geography, and the Vikings who wintered over here left behind no road signs welcoming later visitors to Vinland. Indeed, the name Vinland itself could have been wishful thinking, like all those concrete American towns with sweet pastoral names. For L'Anse aux Meadows is not, to put it mildly, the Rhone Valley or hearty Burgundy.

As I walked away from the site, I saw George Decker's son Lloyd scything a clump of grass. He was a tall wiry man whose face was an elaborate hieroglyph of lines and contours; if each man is responsible for his own face after the age of forty, then Lloyd had done a very good piece of work on his. It was Lloyd who turned over the first clod of soil for Helge Ingstad back in 1961; later he helped excavate the site. I figured he might have squatter's rights on the subject of Vinland, so I asked him if he thought his father's pasturage was the Viking Shangri-La. "Well," he shrugged, leaning on his scythe and gazing down on the berry-flush earth, "it's something . . ."

Eloquent words. I agreed with him entirely and did not pursue the question, as I appreciate having a few mysterious "somethings" lying around just to keep my thoughts on their toes and those toes themselves on the hard, secret ground.

September 24. Evening. I'm curled up in my sleeping bag reading Suetonius's *Lives of the Twelve Caesars* when there comes a tap-tap-tapping on my tent flap. Elderly gentleman slightly in his cups wishes to advise me that I oughtn't to be camped in this neck of the muskeg. Am I violating some municipal ordinance? No, he says, only I happen to be camped close to an Abominable Snowman's lair and hell hath no fury like an aroused Snowman. But the Abominable Snowman is found only in a few remote parts of the Himalayas . . . Ah, that just goes to show the limitations of your Yankee know-how, because we have Abominable Snowmen in Newfoundland, too. William Decker shot one that was ten and a half feet tall in the year of Our Lord 1897. We have banshees and widow's curses, too, he says, but one thing we don't have, I'll tell you, my son, is this: A market for our flounder and halibut. By God, we don't, my son!

September 25. Quite early I ride with Billy and his eldest son Billy Jr. to the fishing grounds off Great Sacred Island. Dawn doesn't crack this time of year, it merely dilutes the night. Thus we get another typically somber day, with pinpricks of light in the tenebrous clouds and the pale duck-egg blue of the distant sky just out of reach beyond Belle Isle. Billy drops his trawl lines from a nineteen-foot dory he constructed of spruce planks after a nor'easter picked up his previous dory (which he had also made himself) and shattered it broadside against Tom Colbourne's fishing stage.

A flock of Canada geese bound for the South hurtles by at spectacular speed, delivering their rising two-note honks in loud falsetto. A few fat idle flakes of snow sift down and then the wind picks up and starts to manufacture hock and tongue from the waves. *Storm?* I shout to Billy. *Storm!* he shouts back, and continues to haul up cod (a bit like hauling up mud) while Billy Jr. baits the empty hook with squid and tosses it in the water, yelling away a tickleace (kittiwake) that's making a grab for the bait. *"Out here we're free, my son,"* Billy shouts more to the wind than to me.

But even freedom has limits and soon we are heading for Great Sacred Island to wait out the brunt of the storm. Mooring lines tied, we clamber ashore on this desolate topknot of rock, wind-scudded bog pools, and the ubiquitous Labrador tea. Great Sacred Island may have been visited by the Vikings and described by them in no uncertain terms as an island wholly devoid of prospects; that was before a storm on November 5, 1947, deposited the *Langlecrag,* a perfect sanctuary, on its shore. We sit hunkered comfortably beneath the boat's aft section and I learn from Billy the reason the Beothuk Indians did not survive in Newfoundland: instead of using their firearms to kill their white adversaries, they dismantled them and used the parts for jewelry. I hear about the grandfather sharks off Cape Bauld who are as old as the sea itself and about a certain caribou that fell in love with one of Jobe Anderson's ewes. And of the best cures for rheumatism (a poultice of liquefied jellyfish), for flatulence (a mixture of milk and soot), and for the itch (flowers of brimstone mixed with lard). Of the Eskimo Magishoe, from nearby Magishoe's Marsh, the last Eskimo in all of Newfoundland, who simply disappeared (stolen by fairies?) one summer's day during the middle of the last century. And of Billy's own Aunt Jessie, who went berry picking fifty years ago and has not been heard from since. It was just possible she'd been stolen by some of those bloody fairies, too . . .

Into October I made camp at L'Anse aux Meadows, reading Suetonius, delighting in the absence of black flies, and listening to Billy Heddison's cures for incurable ailments. The days grew perceptibly shorter even as they seemed to grow brighter, more lustrous. Long sunsets of mauve and smoky blue slowly unfurled above the fissured cliffs of Labrador. Indian winter retreated into an autumn that hung its haze over rock and sea like a benediction. Whole doughnut-bald hillsides metamorphosed to velvety crimson, thanks to the profusion of small vaccinium shrubs and Labrador tea. Ragged V's of Canada geese drifted loosely southward like strands of gossamer in a light breeze. And for once there were indeed light breezes rather than the usual blustery winds.

One day I trekked ten miles inland into the barrens. There was no trail, for in a fisherman's domain only the coast is worth walking upon, and on this landward trek I met with not a single sign of the Ascent of Man except for, inexplicably, a lone tennis shoe lurking like a dull morose fish just beneath the surface of a bog pool. I climbed hummocks of bare rock furrowed by ice and tenanted only by lichen and the delicately composed stigmata of purple saxifrage. From the summit of a steep little hill an old dog fox gazed down on me, its coat shaded out to chestnut and already thickened against the winter. I was picking berries when I was struck by that sixth or perhaps seventh sense that tells a person that something big and alive is close at hand. I looked up. Not twenty feet away was a white moose — an albino bull — taking his ease in a patch of stunted willows, antlers flashing in the midafternoon light. I stared at him until he saw me, and then he quickly trotted off along the hillside. I knew I would never see another white moose again; he was a rare ghost bred by an equally rare October day.

A few days later I watched the blacktop arrive in L'Anse aux Meadows. The whole village turned out for this minor event of major proportions, which was a bit like the arrival of the first mobile home on Foula. Half the people were celebrating the new road with beer and a kind of moonshine made from peas; the other half appeared to think this new road a terrible idea and they were celebrating that, too. After sampling some of the moonshine, which tasted like crankcase oil, I got into my rented vehicle and headed for the outport of Quirpon (French corruption of the English "harpoon"). At the turnoff the road was once again rampant with the sort of potholes I had come to know and love in last places. Potholed roads encourage lost civilities in the soul; in Quirpon a man from whom I asked directions gave me a pair of new hand-knit stockings, saying only that he didn't want them (also, an American sailor had once bought him a girl in a brothel in Halifax, Nova Scotia, and he still felt grateful).

Quirpon was a jumble of plywood houses and fishing stages, a Catholic church and a Pentecostal church, a Salvation Army hall, and a shop, all just out of the sea's reach. Like L'Anse aux Meadows, it had once played host to Maritime Archaic Indians and seasonal French fishermen; now it plays host only to maritime archaic inshore fishermen. It was the sort of place where an infamous Nazi war criminal could disappear forever and no one would ask him any questions except whether he was taking fish with squid or herring bait. I talked to a woman who had never heard of Ronald Reagan and a man who thought the pope lived in Ottawa. Then I began to walk toward Fortune, a place whose name held the possibility of hope but whose fate was just as unfortunate as hundreds of other outports with names like Gripe Point, Famish Gut, Bad Bay, Empty Basket, and Bareneed. Fortune was rendered obsolete one day twenty-five years ago by the resettlement program: its ghosts inhabit the eternal camaraderie of the afterlife with the great auk and the Ungava grizzly.

It was another radiant autumn day. Quirpon Island and Jacques Cartier Island sat brightly in the harbor, their rocky shores glittering with brocades of seaweed. From Noble Cove I climbed a low granitic hill and then a higher granitic hill, walking over ground that was at once rich and barren, like the tundra country of Labrador. The rich part was a sward of mollyfudge (Newfoundland slang for soft, springy moss), yarrow, and berry bushes; the barren part, an ocean of small boulders that stuck up like inquisitive bald heads. At the top of one hill I found a fox's cache, which contained six goose feet and a doll's plastic leg, all pointing in the same direction. Then I descended into a long promenade of alders. Anyone who has ever done hand-to-hand combat with this apostate of the birch family knows why Jupiter turned Phaeton's sisters into alders: no tree is more mean-spirited, its branches springing back with a lash and a snap, nor any jungle more ill-disposed toward ordinary human access. These alders grew in dense clumps alongside a stream; they took on every crooked bend of the water, but I knew from my directions that I was following the right route, a path along that stream. The alders had grown there since the last Fortune-ite hiked this way to purchase fishhooks from the shop in Quirpon, or the last person from Quirpon came this way for a dance in Fortune.

I came out on a narrow tidal flat between two low-slung headlands. For nearly a mile I walked along a rocky beach littered with broken clam shells and whore's eggs (Newfoundland slang for sea urchins). There were ravelments of fisherman's twine and a few rubber gloves, but no bomb casings this time. Flocks of scoters skimmed over the water. A black-backed gull floated overhead and then, observing me, caught the light updrift of air from the cliffs and swiveled off inland. Camel

Island and Broize Point blocked the open water, so the sea was hardly more than a soundless breath swelling from afar, a slow pulse-beat so soft I could hear the tiny sputterings of crustaceans tucked away in the rocks. I leaped over a string of tide pools.

And then I came to Fortune: a dozen or so tumbledown frame houses, blind windows staring, weather-tormented boards, a church, and a collapse of gray planks that might have been a house or an outbuilding, too late to tell now. All of them rested on a grassy slope that cascaded down to the community stage, where I found a number of empty barrels of Jamaican rum. Outporters couldn't afford to buy actual rum; instead they bought the barrels, soaked up the rum residue with water, then ran it off like moonshine.

One or two of the houses looked as if they might survive with a little fixing up, but most were totally beyond repair. I felt less like an invader than an archaeologist. Floorboards were caved in, storage cellars exposed. One roofless house had an old tin stove just like the Mugfords' stove in West Bay, but it had been crushed (a falling beam? an Abominable Snowperson?) almost beyond recognition. In another dwelling a stovepipe hung by a string, tottering eerily back and forth and clanking against the wall when the wind blew. I found a 1961 calendar on the floor, its ancient-looking pages curled up into big russet cocoons. And in a neighboring pile of floorboards I found a piece of a "Home, Sweet Home" sampler — the final Newfie joke, perhaps. The whole north face of the church-school had been shorn off clean as if by a giant chain saw; heaped on its floor were blue books, prayer books, arithmetic books, spelling books, and one dead muskrat. I opened a blue book and read: "Newfundland joined the Confedration of Canada in 1949. It's chief product is fish." And in another, in a child's big block lettering: "Francie lives in a pretty white house by the sea." In another, obviously written for Sunday school:

". . . the one abiding Sin, which the Lord does not forgive, is an adult tree . . ." The rest had been licked into a meaningless blur by rain and the elements.

The one abiding sin of Fortune was being too far away. But too far away from *What?* In Straitsview, the next outport east from L'Anse aux Meadows, I had chatted with a man who asked me where I lived. "Hundreds of miles away," I said, "near Boston." "But how do you manage to live so far away?" he asked, being himself from a place at once fly-specked and the center of the universe.

I was tired. It had been a long trip, not just from Noble Cove, but also from Bergen, Norway, over four months ago. I sat down and rested on a little hogback, among the wild mustard and chickweed, and ate some of the bread I'd brought from Quirpon. Here in this last place, or near it, dawn first touches the New World. But now the gray fingers of dusk had begun to brush the landscape and its houses were dimming, phantoms extinguished. And as I gnawed on the bread and looked down on the derelict houses, I remembered something Billy Heddison had said:

"By God, they had some bloody fine parties in Fortune, my son."

Chapter 14

LANDFALL

I GOT BACK to my hearth in Cambridge, Massachusetts, in late October, shortly after the leaves had lost their exuberance. New England was sere and distracted, settling in for the skeletal vignettes of winter. The first few days I kept waking up and expecting to see a tent stretched gracefully over my head; I was apparently stuck on automatic pilot and wandering my own interior atlas.

People, as an entity, seemed a little disquieting. There were simply too many of them. Where I had come from, they were few, and Nature, at once niggling and virtuoso, wielded the big stick. Here the situation was reversed and Nature was constantly being forced to submit to human expansion: *more* shopping centers, *more* condominiums, *more* clogged arteries of concrete. Everyone's rhythms seemed frazzled and hectic, a series of jump-started hypothalamuses jousting — some successfully, most not — with the demands of urban life. I felt as if I were reentering the earth's atmosphere after being whisked away by a kindly UFO to a retrograde planet. I coughed, snorted, complained, grappled, suffered decompression, and . . . adapted.

One day I was walking along Memorial Drive in Cambridge not far from Mount Auburn Hospital when I happened to notice the plaque honoring Leif Ericson's alleged visit ("On this spot in the year 1000 Leif Erikson built his home in Vine land"). Apparently a nineteenth-century Harvard chemist and gentleman historian named Eben Norton Horsford was so convinced that this "spot" — formerly Gerry's Landing — was Vinland that he paid for the plaque himself and personally saw to its installation.

For some time I'd known about this rather eccentric Viking site on the banks of the Charles River. I had never visited it, mainly because I avoid taking walks where primacy is given to cars and their needs. Especially I avoid taking walks where those cars seem determined to emulate low-flying Labrador aircraft, tearing along at absurd speeds as if they were actually going somewhere. But now here I was, and I cast my well-traveled eye upon this final piece, however unlikely, of Viking terrain. The grass was brown, stunted, and permanently discolored by pollutants. The Charles was sluggish and discolored, though perhaps not permanently, by more pollutants; on its banks sat a couple of two-toned ducks — avian bathtub rings. I couldn't float a fanciful Viking longship on the river because the punts of the Harvard rowing team kept getting in the way. And in the air hung a sort of velvety smog that made breathing a liability. Leif Ericson himself would have managed one quick sniff of this air and then moved on to a more favorable anchorage.

I coughed, snorted, grappled, and — *un*adapted. My mind rebelled against this unhappy scene and I began to plan my next journey. I would require a new pair of hiking boots, for I would be traversing lots of extremely rough country. I would need a new passel of maps and more Ruttledge no. 2 word processors, as well as a phrase book to help me communicate in the obscure local vernacular. Where this place would be was anybody's guess, including my own. I only knew that it would be beyond the last tollbooth on the last scrap of potholed pavement at the very end of the turnpike . . .

ACKNOWLEDGMENTS

The swath of the North Atlantic is so vast and daunting that without the assistance of a good many people and organizations, my journey would not have been nearly so successful. The following airlines picked me up whenever my feet rebelled: SAS Scandinavian Airlines; Icelandair (William Connors, Sigurdur Helgason, Sr.); Greenlandair (Ole Oxholm); Air Nova (Paul Lannon, Bruce MacLellan); Air Canada (Dave Crisman); and Lab Airways. The following tourist organizations were especially helpful, even though they must have been aware of my constitutional dislike of tourism: Danish Tourist Board, New York (David Swindell); Rogaland Tourist Association, Stavanger, Norway (Pelle Nielsen); British Tourist Authority, New York (Bedford Pace); Faeroe Islands Tourist Office, Torshavn, Faeroe (Jakob Veyhe); Iceland Tourist Board, Reykjavik, Iceland (Birgir Thorgilsen); Narssaq Turistkontoret, Narssaq, Greenland (Henrik Grum); Newfoundland-Labrador Department of Tourism and Development (Shirley Letto, Joe Goudie); and the tourism division of the Canadian Consulate in Boston (Janet Aiton, Ralph Johansen). I also received the hospitality of Hotel Hafnia, Torshavn, Faeroe; Royal Inn, Goose Bay, Labrador; Deer Lake Motel, Deer Lake, Newfoundland; and

Shallow Bay Cabins, Cow Head, Newfoundland. The Smyril Steamship Company, in transporting me from Norway to Shetland and from Shetland to Faeroe, treated me with a civility that I had no right to expect. My thanks to Dominic Capezza for including me on one of his Greenland cruises, though all I could offer his passengers in return was an evening of scatological folktales. My thanks also to Peter Andersen, Angmagssalik, East Greenland, who smoothed my entry into a rough-and-tumble but transcendently beautiful part of his island. Thorgeir Thorgeirsson and Vilborg Dagbjartsdottir, Reykjavik, Iceland, retrieved me from the elements so frequently that I came to view their home as my own. Ragnar Ragnar and his charming wife Siga opened the doors of their house to me in Isafjördur and Jóhann Pétursson graciously opened the doors of his lighthouse at Hornbjarg, Iceland. Gladys Anderson, Max Anderson, Job Anderson, and Bill Bartlett were a wealth of information about the tangled skein of lives and genealogies in the Great Northern Peninsula of Newfoundland. Eastern Mountain Sports and the Rockport Company helped to outfit my journey; whatever dissolved, rotted, or fell apart by the end was my fault, not theirs. Richard P. McDonough kindly lent his ears to my filibustering and seldom failed to offer helpful comments. Tess Zimmerman, Stephen Loring, and David Belknap responded to my bibliographic requests — some of which were quite cruel — with unflinching cheerfulness. Finally I would like to express my gratitude to the John Simon Guggenheim Foundation for their generous support during the writing of this book. Well, not finally. *Finally* I'd like to thank the stalwart citizens of these last places, who fed, entertained, and generally intoxicated me.

Chapters from this book have appeared, in slightly altered form, in *Pequod* and *Margin*. "Dog-Day Revolutionary" draws primarily on J. F. Hogan's *The Convict King,* Rhys Davies's *Sea Urchin,* and Thorsteinn Jónsson's *Jorgen Jorgensens Líf.*

LAWRENCE MILLMAN
Cambridge, Massachusetts
January 1989

CPSIA information can be obtained
at www.ICGtesting.com
Printed in the USA
FSOW01n0801140417
33124FS